3 0063 00332 4847

Eastern   Oct 2019

FRIENDS
of the
Davenport Public Library

"Celebrate The Printed Word"
Endowment Fund
provided funds for the
purchase of this item

REPOWERING CITIES

REPOWERING CITIES

Governing Climate Change
Mitigation in New York City,
Los Angeles, and Toronto

Sara Hughes

CORNELL UNIVERSITY PRESS   ITHACA AND LONDON

Copyright © 2019 by Cornell University

All rights reserved. Except for brief quotations in a review, this book, or parts thereof, must not be reproduced in any form without permission in writing from the publisher. For information, address Cornell University Press, Sage House, 512 East State Street, Ithaca, New York 14850. Visit our website at cornellpress.cornell.edu.

First published 2019 by Cornell University Press

Printed in the United States of America

Library of Congress Cataloging-in-Publication Data

Names: Hughes, Sara, 1981– author.
Title: Repowering cities : governing climate change mitigation in New York City, Los Angeles, and Toronto / Sara Hughes.
Description: Ithaca : Cornell University Press, 2019. | Includes bibliographical references and index.
Identifiers: LCCN 2018059406 (print) | LCCN 2019000200 (ebook) | ISBN 9781501740428 (pdf) | ISBN 9781501740435 (epub/mobi) | ISBN 9781501740411 | ISBN 9781501740411 (cloth ; alk. paper)
Subjects: LCSH: Climate change mitigation—Government policy—New York (State)—New York—Case studies. | Climate change mitigation—Government policy—California—Los Angeles—Case studies. | Climate change mitigation—Government policy—Ontario—Toronto—Case studies.
Classification: LCC QC903.2.U6 (ebook) | LCC QC903.2.U6 H84 2019 (print) | DDC 363.738/74561—dc23
LC record available at https://lccn.loc.gov/2018059406

For Jean-Luc, whose support is tremendous in all ways, and Pascal and Josephine, who may one day benefit from our efforts to address climate change

Contents

Introduction: *The Shifting Ambitions and Positions of City Governments*   1

1. Progress or Pipe Dream? *Cities and Climate Change Mitigation*   17
2. Evaluating Urban Governance: *A Three-Part Framework*   46
3. Made to Measure: *Tracing Unique Climate Policy Agendas in New York City, Los Angeles, and Toronto*   79
4. The Means Behind the Methods: *Governing Strategies to Reduce Greenhouse Gas Emissions*   116
5. Are We There Yet? *Identifying and Evaluating Urban Progress on Climate Change Mitigation*   148

Conclusion: *Prospects and Consequences of Repowering Cities*   165

Acknowledgments   181
Notes   185
References   191
Index   209

REPOWERING CITIES

Introduction

THE SHIFTING AMBITIONS AND POSITIONS OF CITY GOVERNMENTS

> **Historians will say the turning point came like a bolt out of the blue. . . . They will say it began in cities, with courageous leaders and far-sighted planners.**
>
> George Heartwell, *former mayor of Grand Rapids, Michigan*
>
> **The fact is that some of the most promising, innovative, effective climate solutions are coming directly from mayors around the world and around the United States.**
>
> John Kerry, *former U.S. secretary of state*

An inversion is taking place in which solutions to many of the world's most pressing problems are being pursued by, and channeled through, cities. Once considered the purveyors of street repairs and sewer mains, city governments are now being heralded as innovative, entrepreneurial, and dynamic actors ready to take on societal challenges that other levels of government seem unprepared or unwilling to address. Once seen as inherently ungovernable (Yates 1977), city governments are now viewed as bastions of democracy and pragmatic problem solvers (Sancton 2011; Barber 2013; Katz and Bradley 2013). Benjamin R. Barber argues in *If Mayors Ruled the World*, "if we are to be rescued, the city rather than the nation-state must be the agent of change" (Barber 2013, 4). The notion that city governments are the world's problem solvers is taking root.

City governments are viewed, and are viewing themselves, as able to effectively pursue major policy agendas once considered the sole purview of national governments. From labor to immigration to climate change there has been a shift in both practice and rhetoric to cities. In the United States, city governments from Bangor, Maine, to Los Angeles, California, are raising the minimum wage for their residents, even as many state governments scramble to prevent them from doing so. The New Localism movement in Europe has initiated a resurgence of community-oriented policies that seek to harness innovation in cities and develop new, locally led governing arrangements for national growth and development goals. The rise of paradiplomacy by city governments—led by São Paulo in Brazil—is reshaping global dynamics of trade and investment. Mayors are even taking on structural social inequality, with Boston and Detroit both experimenting with free community college for low-income high school graduates.

In this book I focus on local efforts to address global climate change. I examine the means by which city governments pursue climate change mitigation, or reducing the greenhouse gas (GHG) emissions produced by urban systems, and to what ends. My primary interest is not explaining cities' motivations or delineating the many challenges city governments face as they work to reduce GHG emissions, although I do discuss each of these in the next chapter. Rather, I use New York City, Los Angeles, and Toronto as empirical cases for examining how city governments move from making a commitment to climate change mitigation to fulfilling it: the governance processes and political mechanisms by which a city government purposefully and intentionally works to steer its city in a new direction. I also take the important step of evaluating the direct and indirect effects of these efforts. The book provides deeper insight into the role city governments play in urban governance, and challenges emerging models of federalism that privilege localism and promote urban pragmatism. I use the cases to better understand how and whether city governments can effectively govern GHG emissions and, by extension, whether local climate change leadership is a substitute for national or international policy. I also identify promising ways forward for the many city governments seeking to make progress on the climate change mitigation goals they have set for themselves.

Renewed faith in cities stems from a view of city governments as pragmatic, responsive, relatively nimble bodies of government. The loudest champions of this perspective emphasize the nonpartisan nature of local politics and the propensity of local decision makers to focus on action and realized outcomes rather than the demands of special interests. Unlike national bodies, the argument goes, city governments are free from partisan bickering and therefore able to move forward with good ideas as they arise. The local nature of city government is itself seen to be an advantage, as they are more attuned and responsive to community needs and preferences. City governments are viewed as more nimble, less bureaucratic, and able to experiment with new ways of doing things that other levels of government may find difficult or cumbersome. In this sense some of the promise of cities lies in their ability to serve as a test bed for other levels of government. Perhaps most centrally, many cities are demonstrating a level of political will and leadership often absent at other levels and have a growing number of donors, agencies, and nongovernmental organizations—not to mention residents—with a real stake in their success.

There are good reasons to push back against this optimistic portrayal of city governments: while they may be responsible for keeping the lights on, how they choose to do so, and whose interests are met in the process, is shaped by a host of political, economic, and institutional forces. Cities face strong incentives to prioritize economic development and are vulnerable to capture by growth-oriented coalitions (Molotch 1976). Many cities seeking to lead on progressive issues like climate change do not have particularly strong environmental track records and face serious issues of rising social inequality. City governments are notoriously resource constrained, which is particularly problematic when confronting new and challenging issues. Climate change mitigation is also one of our most wicked policy problems, one steeped in complexity and uncertainty. Reducing GHG emissions implies deep and long-lasting changes to cities as we know them and are likely to require fundamental urban transformations.

The current urban inversion can also be seen as the culmination of the neoliberal political projects of decentralization and state retrenchment, particularly in the United States and Canada. In Canada, neoliberal reforms initiated in the 1990s have restructured social services and reduced the role of the federal government in urban investment and policymaking (Hackworth and Moriah 2006; Dunn 2008). A similar dynamic has played out in the United States, as Republican and Democratic administrations have both worked since the 1980s to pull back from the New Deal consensus, shifting responsibility for urban fortunes from federal policy and investment to local decision makers and stakeholders (Weaver 2016). Progress on federal environmental policy has been particularly stymied in the U.S. due to growing polarization and congressional gridlock, prompting advocates to seek out alternative venues, including cities (Klyza and Sousa 2008).

While there may be reasons for mayors to rule the world, it is not always clear that they rule their cities. The rise of cities, and the tensions and contradictions it makes visible, opens up new and important questions about how change happens in cities, the place of city governments in these change processes, and where cities may ultimately be headed if they do indeed take on the role of the world's problem solvers. It compels us to better understand the capacities and tools city governments have for governing complex problems and how they can be put to their best uses. We must ask how city governments can effectively pursue solutions to complex social and environmental challenges given the institutional and resource constraints they face; how these processes play out differently in different cities; how cities might ultimately change as a result of these shifts; and, ultimately, whether we are wise to put our faith in cities.

Climate change mitigation is one of the most prominent and urgent examples of the urban inversion. In many ways a city-led response to climate change is unintuitive: reducing GHG emissions in a meaningful way is in many ways a collective action problem, as reducing GHGs on a scale that will affect the global climate requires action beyond the scope of any single city—or nation. Cities face strong incentives to free ride on the mitigation efforts of others.

Despite these incentives, over the last twenty years, hundreds and now thousands of city governments have made commitments to reduce their GHG emissions in an effort to avoid the worst consequences of climate change, even as climate policy has largely stagnated at the national and international levels. Cities are setting increasingly ambitious GHG reduction targets even as many state, provincial, and federal governments have failed to take meaningful action. The ambition and enthusiasm of city governments has earned them official status in international negotiations and shifted the center of gravity away from the traditional multilateral process (Hoffmann 2011). In recognition of this growing sense of local leadership and potential, city governments were granted official participant status at the December 2015 UNFCCC 21st Conference of the Parties in Paris for the first time since the agreement came into force in 1991, and city leaders successfully lobbied the French government to include a "Cities and Local Governments Day" as part of the talks. The mayors of London and New York City have committed to divest city pension funds from fossil fuels and are encouraging others to do the same. Cities have become a critical lynchpin in the locally mobilized, polycentric climate governance system that is emerging.

Cities' contributions to climate change solutions are exciting: the ambitions of city governments have reinvigorated the climate change community and have led many to speculate that climate change will be solved in cities, or not at all. The leadership and ambition of city governments to tackle climate change has injected a new sense of hope for a climate change response.

What is too often left out of discussions of the transformative and rule-bending potential of cities is the formidable challenge cities face in turning their ambitions into reality. The work of meeting a GHG reduction target intersects with the realities of governing a shifting, complex city that is likely on a very carbon-intensive trajectory. Reducing a city's greenhouse gas emissions by 80 percent, 50 percent, or even 30 percent requires fundamental changes to take place: energy use patterns, transportation and mobility networks, electricity generation, and waste management strategies all stand to be restructured. The infrastructure and behavior patterns these services shape and reflect are the foundation of urban economies and communities. Climate change mitigation is emblematic of what is called a "wicked" problem (Koppenjan and Klijn 2004). It is steeped in complexity, and uncertainty reigns when it comes to our understanding of the interdependent drivers of urban GHG emissions and our ability to pursue effective and equitable solutions.

Despite these challenges, more cities continue to set more ambitious climate change goals. Transnational city networks, such as the C40 Climate Leadership

Group and Local Governments for Sustainability's (known commonly as ICLEI) Green Climate Cities program, now represent over 1,000 city members; in 2010 nearly two hundred city governments from forty-five countries signed the "Mexico City Pact," pledging to reduce their cities' GHG emissions (World Mayors Council on Climate Change 2010b). Cities from the global North and South have set ambitious targets to reduce their GHG emissions by as much as 80 percent over the next several decades. Given the uncertain and shifting international response to climate change, and the rapid acceleration of global GHG emissions, their ability to succeed may be crucial.

Cities have made their intentions clear; less clear is how cities are pursuing the policies and strategies that will allow them to make good on these commitments. While significant scholarly attention has been given to the question of *why* city governments are committing to address climate change, and to the obstacles and barriers they face, much less has been given to the question of *how* cities can or do proceed in governing GHG emissions.

The Framework

Repowering cities refers to the dual challenge presented by urban climate change mitigation of reconfiguring the systems that power our cities while harnessing the governing powers of city governments in the pursuit of transformative change. In one sense, repowering cities refers to the challenge of reconfiguring and reorienting the infrastructures, behaviors, institutions, and economies that power our cities in order to reduce their GHG emissions and contributions to global climate change. Meeting urban climate change mitigation goals requires that cities be "powered" in new ways: using renewable energy supplies, alternative modes of transportation, and an energy-efficient built environment.

A commitment to repowering cities in this way is not just useful but *necessary* for any serious attempt to reduce global GHG emissions. City governments have authority over a significant portion of GHG emissions by virtue of their role in shaping energy systems, transportation, land use planning, building regulation, and waste management. Even in Canada, where city governments are not typically considered to hold significant authority, they have direct or indirect control over 44 percent of the nation's GHG emissions (Federation of Canadian Municipalities 2009). Any serious attempt to slow the rapid growth of global GHG emissions, and therefore avoid the worst effects of climate change, will require significant changes in cities and a willingness on the part of city leaders to rise to the challenge. It requires that cities be fundamentally reconfigured, with implications for energy and transportation systems, behavior patterns, and lifestyle choices. It requires that city residents interact with existing services and infrastructures in a new way and toward new ends.

The stakes of climate change are high—the future of our communities and economies are deeply implicated. Therefore the question of how to govern GHG emissions in the city moves quickly from interesting theoretical territory to the realm of the immensely practical. If city governments are to take on the task of responding to climate change, a much clearer road map is needed. We are only beginning to understand the challenge of reducing GHG emissions in cities and what implications this work might have for urban infrastructure, mobility, equality, and economies (Marcotullio et al. 2014; Romero-Lankao et al. 2014; Bulkeley 2015). Some of the most pressing questions surrounding climate change, therefore, hinge on whether and how cities can be powered in entirely new ways in pursuit of deep GHG emission reductions and the metrics by which we should evaluate success.

In another, broader sense, repowering cities refers to the governance challenge climate change mitigation poses for city governments and the subsequent need to revisit and reassert the power cities have to take up these challenges. While the ambitions of city governments are in many ways a good news story for global climate change policy, governing GHG emissions is an entirely new endeavor. As a "wicked" policy problem, it is steeped in strategic, institutional, and substantive uncertainty (Koppenjan and Klijn 2004). Developing the policies, programs, and partnerships necessary to influence urban GHG emissions requires city governments to stretch and expand the limits of their authority and chart new political, institutional, and technical territory.

As a result, it is not entirely clear that city governments are well positioned to take up the mantle of climate change leaders. Skeptics point out that city governments are likely to encounter important limits to their jurisdiction as they seek to reduce GHG emissions. For example, regional energy and transportation systems (often major contributors to urban GHG emissions) may be controlled privately or by higher levels of government. City governments are embedded in broader political–economic systems that can determine the viability of certain policy solutions or availability of willing climate policy partners. Tracking and monitoring GHG emissions is a technically complex task, and city governments may not have access to the skills and datasets necessary to develop and implement effective mitigation policies. Perhaps most importantly, efforts to reduce GHG emissions will compete with other local policy priorities for financial resources in a notoriously resource-constrained environment.

Effectively governing GHG emissions will require that city governments are repowered and recentered. They must find creative ways of mobilizing and directing a vast array of people, investments, and activities toward a new, ambitious collective goal. City governments must act both independently and in partnership, leveraging multiple sources of authority, and using a variety of tools and strategies as they work to take their city in a new direction. For cities to take on climate change mitigation, they will need to do more than implement a new policy or launch a new program; they will need to steer their cities in fundamentally new directions.

Repowering cities refers to the means by which city governments can and do govern toward a fundamentally new aim like climate change mitigation. Throughout the book I seek to clarify the role of city governments in governing complex policy challenges; understand the dynamics at play as they pursue a change of course for their cities; and make explicit, through both theoretical development and empirical analysis, the role and space for action on the part of city governments and how their efforts can best be utilized. Rather than focusing on the factors or characteristics of cities that might make them more or less likely to adopt a policy or make a particular commitment, I am interested in understanding what city governments are doing, can do, and must do to steer their cities in new directions.

The ambitions of city governments have prompted an emerging discourse of urban leadership, but we lack tools for examining the processes engaged and outcomes achieved as cities work to make progress on climate change mitigation. I develop in the book a framework for understanding how and whether cities are being repowered. First, I clarify which dimensions will be city-specific and which are likely to be universal across cities. I distinguish between the choices city governments make about *policy agendas* for reducing GHG emissions and the *governing strategies* they use to mobilize the resources and actors needed for change. I argue that choices about policy agendas are driven by the opportunities and barriers that particular cities face. While much of the existing research on cities and climate change uses a city's context to predict whether a city will act in the first place, I argue that the more important role of a city's context is in shaping the choices they make about *how* to pursue their climate change goals. In terms of policy agendas, repowering cities plays out in unique ways.

There are also processes that play out in very similar ways in different cities. The second element of my framework is the identification of shared governing strategies for mobilizing the resources and engaging the actors needed to redirect urban practices and policies and meet ambitious goals like reducing GHG emissions. I argue that regardless of the material or political context of cities, urban climate governance is a function of the same set of strategies: institution building, coalition building, and capacity building. These governing strategies serve to reduce sources of uncertainty inherent in complex problems and reorient a city's trajectory. They enable learning and long-term commitment, coordinate political and financial resources, and facilitate behavior change. Repowering cities requires more than a goal and a policy toolkit; for city governments to take up complex problems like climate change, more fundamental strategies are required to build a governance infrastructure supportive of change.

Third, I clarify valuable metrics for evaluating the outcomes and consequences of climate mitigation efforts. First is in changes or reductions in GHG emissions, and especially reductions at a rate consistent with the city's longer-term goals. If city governments are governing effectively, we should observe such reductions—to an extent. A challenge with this metric is that it can be difficult to attribute changes in a city's GHG emissions directly to the actions of the city government, and urban GHG emissions can be influenced by policies and programs developed at other levels of government. Further, we often lack adequate records and data to make such claims. The second way to observe the effects of urban climate governance is through broader changes within the city and catalytic effects beyond the city. Within the city this can mean changes to city government culture or practices, new norms of governance and accountability, new relationships among stakeholders and residents, or new political alliances or cleavages. The work of cities also extends beyond their boundaries. The experiences and capacities cities build as they take on an issue like climate change can inform or enable decision making in other cities or other levels of government. As cities become more embedded in the international arena, their ideas and lessons learned can travel very quickly. State and federal governments may look to cities for best practices or new ideas. These ripple effects of urban climate governance can be just as valuable as direct measures of effectiveness.

In developing and applying this framework, my aim is both analytical and normative. Analytically, I seek to specify the role of city governments in urban governance and the means by which they enact this role in pursuit of GHG emission reductions. I presuppose commitment to addressing climate change, and focus on what city governments do to pursue a climate change mitigation goal once it is laid out. Normatively, I am concerned with the effectiveness of urban climate change governance, and seek to identify strategies and approaches to repowering cities that will facilitate a swift reduction in GHG emissions. I argue that if city governments are to make good on their ambitions for climate change mitigation, they need to take up and master new ways of governing and repowering both their cities and themselves.

Research Design and Methods

I use the experiences of three North American cities with climate change mitigation to further develop and apply my framework: New York City, Los Angeles, and Toronto. These three cities made equally ambitious pledges in 2007 to reduce GHG emissions and contribute to global climate change solutions, and they have continued to commit to and update this pledge in the years since. Like a growing proportion of cities, and especially of large cities, they have decided to take up the task of climate change mitigation. Unlike most cities, they have been pursuing these goals for more than ten years, providing an ideal test bed for understanding the mechanisms and consequences of urban climate governance. They also provide a valuable opportunity to evaluate outcomes—how much GHG emissions have changed since the city's work began and what broader effects this work is having on the city and beyond.

New York City, Los Angeles, and Toronto are similar in many ways. They are the three largest cities in North America, bringing with them global city status, and diverse and growing populations and economies. They are located in the two most climate change–reticent developed countries in the world and so are emblematic of locally led ambition to reduce GHG emissions. Compared to cities in other parts of the world, they are situated in similar political systems: representative democracies. They are members of the C40 Cities Climate Leadership Group and have played leadership roles in this and other city climate change networks. Today, they are all working on a goal of reducing their city's GHG emissions 80 percent by 2050.

Importantly, beneath their surface characteristics these three cities navigate very different terrains as they pursue their climate change mitigation goals. Their sources of GHG emissions; levels of authority, capacity, and political leadership; and broader political–economic contexts vary considerably and have changed over time (see figure I.1). They bring very different strengths and weaknesses to the task of climate change mitigation, and they face unique obstacles and opportunities for action. This set of cases allows me to apply and test the proposition that there are unique agendas and shared strategies available for repowering cities. The cases also provide an opportunity to learn whether city governments with these differences seem equally capable of putting their cities on fundamentally new paths—whether they can govern their cities in ways that make significant contributions to climate change mitigation.

As is outlined in chapter 3 in greater detail, these three cities are different in ways that are valuable to my framework. They have been pursuing their GHG emission reduction goals in unique settings: they have different sources of GHG emissions, different levels of authority over these emissions, bring different capacities to the task of reducing GHG emissions, and are positioned in different political–economic contexts. This variation allows me to examine the distinct routes the cities are taking to climate mitigation—routes that I show are shaped by their context—and the extent to which these different cities are all drawing on the shared governing strategies of institution building, coalition building, and capacity building in their work.

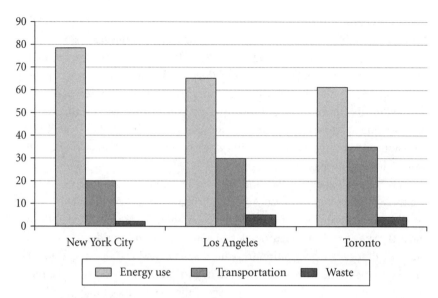

FIGURE I.1 Contribution of energy, transportation, and waste to citywide GHG emissions in New York City (2007 inventory), Los Angeles (revised 1990 baseline), and Toronto (2007 inventory)

I conducted forty-eight interviews with a range of decision makers and stakeholders, city officials, nongovernmental organizations (NGOs), and utility representatives in the three cities between June 2015 and January 2017. These interviews include a former mayor; city councilors; people with prominent positions in mayoral administrations, city councils, and key city agencies; and stakeholders centrally involved in decision-making processes and program implementation, among others. These interviews reveal the motivations and strategies of the three city governments and what they saw as central to their efforts to address climate change. I also draw on government reports, policy and program documents, and newspaper articles to understand the cities' planning frameworks, evaluate progress, and highlight important political moments. Together, these data allow me to provide a rich account of how the three city governments have worked to govern GHG emissions in their cities and contribute to climate change solutions, and evaluate the consequences of these efforts. This systematic approach provides a means for assessing and interpreting their choices and leveraging their experiences to improve practice in other places.

The climate change mitigation story in each city is complex, both technically and politically, and stretches for a decade or more. The empirical chapters serve to examine the political and institutional dynamics of repowering cities rather than provide comprehensive accounts of what each city has done to try and reduce GHG emissions. Subsequently, when tracing the policy agendas the cities have chosen for climate change mitigation and the governing strategies the cities have relied upon, I present the important contours and most illustrative examples. Similarly, when describing a city's energy system or sources of GHG emissions I provide the key elements rather than an exhaustive inventory.

New York City, Los Angeles, and Toronto provide an initial opportunity to understand the process and potential of repowering cities. These three cities are large, global cities with generally strong economies and relatively high capacities. They have made clear commitments to address climate change. These features provide insight into how cities govern GHG emissions and what urban climate governance looks like in cities where it is most likely to be making a difference. As they have ten or more years of experience, we can draw on their successes and challenges to learn more about what it takes to create cities that fulfill a vision of a low carbon urban future. This choice of cases also comes with some limitations. The cases do not account for the full range of urban variation in the United States and Canada, let alone the range of conditions found in cities beyond this region. Nonetheless, I use these cases to take a much-needed first step toward better understanding the role of city governments in governing the complex challenge of climate change and the opportunities available for repowering our cities.

Key Contributions

I make three main contributions with this book. The first contribution is a reexamination and specification of the role of city governments in contemporary urban governance settings. Scholars of urban politics and governance have long emphasized the inherent limits of city governments in governing the city (Peterson 1981; Stone 1989; Pierre 2011). Those governing the city have often been understood to have a necessarily limited scope of authority. But city governments have consistently demonstrated an ability to shape their city's trajectory even in the face of powerful, complex, and global forces (Clarke and Gaile 1998; Savitch and Kantor 2002; Sellers 2002a). Indeed, many urban governance scholars have maintained the importance of city governments to urban development trajectories (Stoker 1998b; Savitch and Kantor 2002; Pierre and Peters 2012; Rich and Stoker 2015). The role and strategies of city governments in domains beyond economic development is less understood.

The dominant frameworks of urban regime theory and urban governance provide excellent starting points for conceptualizing the task of governing a city and positioning city governments within the urban milieu. Urban regime theory gives us the "iron law," that "for any governing arrangement to sustain itself, resources must be commensurate with the agenda pursued" (Stone 2015). The ability of city governments to pursue an agenda is contingent on their ability to mobilize the necessary resources and actors (Stoker 1995). Urban governance emphasizes the diversity of partners implicated in urban agendas and the range of roles that city governments can play in subsequent governing relationships (Pierre 2011, 2014).

Ultimately, neither specifies the means by which city governments lead and pursue an agenda as transformative as significantly reducing urban GHG emissions. In emphasizing so heavily the limits of city governments, we have largely failed to theorize how city governments might pursue a transformative agenda for their city and how they can lead on an agenda like climate change mitigation.

The rise of climate change on urban agendas therefore requires reexamining the role of city governments in shaping and directing the city; local climate change leadership makes clear the shifting priorities and roles of city governments. As a collective aim, climate change mitigation requires a more central role for government and public authority within urban coalitions and the creative use of such authority for mobilizing a dizzying array of necessary resources and actors (Bättig and Bernauer 2009; Heitzig, Lessmann, and Zou 2011). I argue that city governments can and do use coalition building as a tool for achieving their climate change goals. But repowering cities also extends beyond coalition building; it requires institution and capacity building in the service of reducing the uncertainties inherent in climate change mitigation and further mobilizing the resources and participants needed for change.

Climate change mitigation places new demands on city governments—decision makers are "building the plane as they are flying the plane."[1] Cities face uncertainty about how residents or the business sector will respond to new initiatives; uncertainty about their sources of GHG emissions and effective strategies for tackling them; and uncertainty about how to build bridges across sectors, change decision-making practices, and coordinate with other levels of government. It also makes new demands of residents and stakeholders, whose behaviors and priorities must also shift. If we are to understand the potential of city governments to respond to climate change, we first need to grapple with their agency in this complex realm. The analysis presented here takes a step in this direction.

The second major contribution of this book is to shift the empirical and theoretical focus of research on cities and climate change from the challenges and barriers cities face onto the means by which city governments work to overcome them. Previously, research in this area has tended to focus on two dimensions: the characteristics of cities that are associated with particular climate change outcomes (e.g., Krause 2011a, 2012; Homsy and Warner 2015; Ryan 2015) and the barriers cities face in pursuing their climate change goals (e.g., Betsill 2001;

Robinson and Gore 2005; Tozer 2013; Aylett 2014). Likewise, scholarship on cities and climate change has tended to focus on questions of why cities are acting (e.g., Lubell, Feiock, and Handy 2009; Koski 2010; Krause 2011a; Homsy and Warner 2015) or the barriers they encounter (Robinson and Gore 2005; Tozer 2013) rather than how cities might overcome these barriers and realize the benefits of climate change mitigation that motivated them in the first place. Scholars highlighting the inherently political nature of climate change mitigation similarly stop short of charting a way forward or developing a coherent theoretical framework for the task of significantly reducing urban GHG emissions (e.g., Bulkeley and Betsill 2003; Aylett 2010; Bulkeley et al. 2013; Castán Broto 2017). This literature also tends to highlight the important work being done outside of city government to reduce urban GHG emissions (Bulkeley, Castán Broto, and Edwards 2014), which, while valuable, leaves open the question of how city governments leverage their authorities to govern.

This previous research provides a useful baseline for identifying the "factors that matter" for cities and climate change (Ryan 2015). It leaves the door open for greater empirical understanding of the climate change governance process in cities, how city governments might (or have) overcome the barriers that exist, and whether there is a discernible relationship between the characteristics of a city or its context and the choices the city government makes about climate change mitigation. Our understanding of cities and climate change would benefit from greater precision and depth of analysis of the climate change mitigation process in cities. There is also significant work to be done in building theories of urban climate change mitigation that go beyond the structuring effects of institutional arrangements or local economies to account for the role of internal political processes and specify the means by which city governments pursue their goals (Bulkeley 2010; Castán Broto 2017).

Toward these ends, I provide a deep dive into the experiences of three city governments with climate change mitigation. Rather than focusing on broad patterns and correlations between certain elements and outcomes, I discern the steps city governments take to move from a commitment to climate change to tangible outcomes. The empirical and theoretical focus of the book is therefore the processes and mechanisms that ultimately shape the relationship between context and outcomes, allowing us to better account for similar outcomes in very different cities and vice versa. Unpacking the process of governing GHG emissions in cities provides a stronger empirical basis for further theory building and refinement.

An important insight from this work is that the governing strategies available to and used by city governments to reduce urban GHG emissions allow cities to work beyond the legal boundaries of their authority. These institutions do not define the scope of city action on climate change mitigation. City governments use a range of governing strategies, such as enabling or incentivizing the actions of others, which mobilize critical actors and resources that lie outside their formal jurisdiction. This challenges the increasingly popular notion that rearranging legal institutions and authorities is needed to repower cities. The strength of city governments and their capacity to govern is a product of their ability to serve as a catalyst and mobilizer of the disparate actors, organizations, and resources that constitute urban systems.

Finally, this comparative study provides a framework that directly informs practice. For many cities the important question is *how*, not whether, to pursue climate change mitigation. While understanding the relationship between city size or revenues and climate change policy outcomes is theoretically useful, city governments do not have the ability to change these features in a meaningful way. Similarly, there is a growing set of policy tools and best practices available to cities, but they are rarely accompanied by similarly well-developed insights into how such tools and practices can be adopted or implemented (Clean Air Partnership 2007; ARUP 2014).

My argument is that climate change mitigation in cities is underpinned by a common set of *governing strategies* rather than any particular city characteristics or policy agenda. I posit that institution building, coalition building, and capacity building are the foundation of any effort to repower cities regardless of whether it is in service of increasing solar energy generation in the city or launching an energy conservation program. These are also foundational regardless of whether the city is large, has significant authority over its GHG emissions, or has strong political support from City Hall. As cities continue to engage in learning networks and dialogues such as the C40, sharing best practices in process and strategy should be given greater priority.

Organization of the Book

In the chapters that follow, I further develop my framework and present the empirical findings of the case study research. In chapter 1 I provide an overview of current research on cities and climate change, including our understanding of the motivations of city governments for taking on climate change mitigation and the factors and barriers that might shape their success. I emphasize the need to better understand how city governments interact with and overcome these factors and barriers in order to lead on climate change.

In chapter 2 I develop in greater detail my framework for understanding and evaluating the tools available to and deployed by city governments for governing, foregrounding the "how" of urban climate change mitigation. Drawing largely on the urban regime and urban governance literatures, I incorporate the "iron law" of resource mobilization for activating the capacity of city governments to govern (Stone 2015). I outline the special characteristics of the climate change challenge and now it forces us to think in new ways about the role of city governments in urban governance processes. I argue that city governments tailor their climate change mitigation policy agendas to their context, that there are three governing strategies that allow city governments to activate their capacity to govern by mobilizing needed actors and resources, and that our assessment of the contributions of cities to climate change mitigation should include but go beyond direct reductions in GHG emissions to incorporate indirect, catalytic effects at other scales.

In chapter 3 I describe the unique contexts that New York City, Los Angeles, and Toronto confront as they work to reduce GHG emissions. I elaborate the relationship between this context and the choices the city governments have made about their policy agendas: the sectors to target in their mitigation work and the modes of governing they have employed in the process. The cities have developed climate change mitigation agendas that capitalize on their unique opportunities to affect change. While there are some signs of policy transfer and learning, in the main each city has a unique and distinct portfolio of climate change mitigation policies and programs.

I then track in chapter 4 the use and importance of the three governing strategies: institution building, coalition building, and capacity building. The three cities are using these strategies to overcome uncertainty and complexity and fuel their climate change mitigation efforts. These strategies have manifested in different ways: while New York City and Toronto have focused on building stakeholder coalitions invested in and informing city government programs, Los Angeles has focused more on mobilizing voters willing to support ballot initiatives. These governing strategies underpin successful implementation and are often developed on an ad hoc basis as the cities encounter political resistance, capacity gaps, or institutional roadblocks.

In chapter 5, I evaluate the progress the cities have made in reducing their GHG emissions and the effects their efforts have had beyond these direct outcomes. As expected, tracking city-scale GHG emissions is difficult and reporting in many cases is inconsistent. In all three cases, city figures show that emissions are declining and at a rate that is largely consistent with longer-term goals—between 12 percent and 26 percent below baseline levels. Further, as the three cities work to reduce GHG emissions they are having much broader effects on local political and administrative arrangements, other cities' efforts to govern GHG emissions, and the decisions of state and provincial governments. These catalytic effects underscore the important mobilizing role cities play in global climate change mitigation efforts.

Finally, in the conclusion I synthesize the book's major findings and highlight the remaining challenges and tradeoffs inherent in today's locally led climate change agenda. While the cities have made demonstrable progress on reducing GHG emissions, they are now facing the challenge of scaling up their efforts as new targets for 2030 loom. I discuss ways the cities can and are using the governing strategies to do this: by building participatory decision making institutions, building capacity for climate "smart" governance, and expanding and stabilizing the coalition for climate change mitigation. The need for a "big tent" approach to climate change mitigation to make the citywide changes necessary for reducing GHG emissions 80 percent increases the complexity of interests and challenges of coordination. The scope of a viable urban climate change coalition may ultimately set the limits of a locally led mitigation agenda.

Repowering cities does not just change the role of city governments; it has the potential to change cities as we know them in fundamental ways. This framework for understanding the dimensions of urban governance of complex policy challenges has applications in other areas, such as addressing social inequality or the challenges of deindustrialization. The increasing ambitions and responsibilities of city governments should be supported and celebrated, but we must also take seriously the political tensions and tradeoffs they make visible. As city governments continue to reconfigure their role in urban governance along with the built environment, it is critical that we maintain a clear understanding of how these changes take place, for whom, and to what effect.

1
PROGRESS OR PIPE DREAM?
Cities and Climate Change Mitigation

City governments have emerged as leaders in global climate change policy. While international and national climate change responses remain uncertain, a growing number of city governments have taken it upon themselves to set ambitious greenhouse gas (GHG) reduction targets. These targets have often been accompanied by participation in transnational city networks, such as ICLEI's Green Climate Cities[1] and the C40 Cities Climate Leadership Group, but they are typically motivated by more local concerns such as air quality improvements or opportunities for political leadership. Collectively, the ambitions and commitments of city leaders have shifted global discourse and earned them official status in international climate change negotiations. It is clear that cities are having a profound impact on the climate change debate.

This chapter examines how and why this shift has occurred. But more importantly, it raises the question of whether and how this shift can lead to meaningful, locally led reductions in GHG emissions. Urban climate change mitigation is a wicked problem, one steeped in complexity and uncertainty. Unsurprisingly then, the empirical evidence we do have of progress in urban mitigation shows an uneven landscape of successes and failures. While we have made good strides in understanding *why* cities are acting and the kinds of cities that seem to be making more progress than others, we know much less about *how* city governments go from climate change ambitions to reshaping the urban landscape in their image.

The ambitions and leadership of cities are generating optimism for climate change solutions despite political swings and recalcitrance at other levels of government. Without an understanding of whether and how city governments can make good on these commitments, such optimism remains largely unfounded. We must go beyond assessments of the motivations and challenges cities face in their work to reduce GHG emissions and give greater attention to the processes and strategies that repower cities.

The Climate Change Challenge

The increasing intensity of human activity since the 1950s has led to an unprecedented release of GHG emissions into the atmosphere and destabilization of the global climate. Greenhouse gases are a class of compounds that includes carbon dioxide (CO_2), methane, nitrous oxide, water, and ozone. These GHGs—and especially CO_2—are an important regulator of the earth's climate, as they are responsible for absorbing and emitting energy in ways that have historically maintained temperatures hospitable to life on our planet (Harris and Burch 2014). Given current and potential future levels of GHG emissions, the Intergovernmental Panel on Climate Change (or IPCC, a nonpartisan and Nobel Prize–winning body tasked with assessing the state of knowledge on climate change) estimates that global temperatures are likely to rise by at least 1.5°C, and perhaps by as much as 4°C, by the year 2100 (IPCC 2014a).

The IPCC is now unequivocal in its assessment that these rising temperatures and the climatic changes they bring are due to human activity, writing that "human influence on the climate system is clear, and recent anthropogenic emissions of greenhouse gases are the highest in history" (IPCC 2013). The global concentration of CO_2 in the atmosphere has risen sharply, from around 320 ppm in 1950 to 400 ppm in 2016, largely driven by human activity. The majority of anthropogenic CO_2 emissions (65 percent) arises from the burning of fossil fuels (e.g., coal, gasoline, and oil) for electricity generation, industrial processes, and transportation; clearing forests for agriculture and urban expansion accounts for 11 percent, and methane emissions from natural gas extraction, livestock, and waste decomposition account for 16 percent (IPCC 2014b). Globally, and in nearly every region of the world, anthropogenic CO_2 emissions continue to rise.

Given past and current levels of GHG emissions, some level of future climate alteration is inescapable. Climate change is already having an impact in many parts of the world (IPCC 2014a), and the increasing concentration of CO_2 in Earth's atmosphere is poised to further change our climate in fundamentally disruptive ways. In North America, water resources have become more unreliable due to melting snow and ice, wildfires are burning with greater intensity, and terrestrial ecosystems are shifting their geographic ranges and seasonal dynamics (Walther et al. 2002; Stewart, Cayan, and Dettinger 2004; Westerling et al. 2006).

Under scenarios of stable or increasing CO_2 emissions, future risks from climate change for North America are predicted to be significant and include coastal and inland flooding, extreme weather events, extreme heat, and widespread ecosystem disruption (Melillo, Richmond, and Yohe 2014).

Cities are particularly vulnerable to the impacts of climate change due to their concentration of people, economies, and infrastructure. With over half of the world's population living in cities, and an increasingly central role for cities in regional, national, and global economies, the financial and social costs of climate change are heightened in urban areas. Cities will feel the effects of climate change in terms of sea level rise, damage to built infrastructure from storms and heat extremes, health problems for urban populations, greater energy use for heating and cooling, and changing availability of water resources (Hunt and Watkiss 2011). Climate change will also exacerbate existing challenges and vulnerabilities (Bulkeley 2013, 18–19). For example, cities in areas already experiencing water stress, such as those in the western United States, will be most affected by changes to water resources. Populations within cities that are most vulnerable to extreme heat, such as the poor and the elderly, will be most affected by extended heat waves. Such effects are unlikely to be experienced equally by all cities, or by different populations within cities.

Due to their density, cities have lower GHG emission intensities than other places: around 7 tons per capita, compared to the US national average of 21.5 tons per capita (U.S. Environmental Protection Agency 2016), and the average for large North American cities of 9 tons per capita (ARUP 2014). Nonetheless, in aggregate cities contribute significantly to global GHG emissions. The Stern Review on the Economics of Climate Change attributes as much as 78 percent of global carbon emissions from human activities to cities (Stern 2006).[2] In North America, urban activities account for 80 percent of anthropogenic carbon emissions even though cities occupy just 5 percent of the region's land (Jones and Kammen 2014; Seto et al. 2014). For perspective, the World Bank estimates that if the top fifty emitting cities globally were counted as a single country, it would be the third highest emitting country behind China and the United States; the members of the C40 Cities Climate Leadership Group would alone comprise the fourth highest emitting country (World Bank 2010, 18).

While cities in the aggregate are large contributors of GHG emissions, there are often stark differences in GHG emissions between and within cities. For example, the World Bank finds that "richer cities, less dense cities, and cities that depend predominantly on coal to produce energy all emit more greenhouse gases" (World Bank 2010, 16). The characteristics of a city's residents matter—their number, density, income, age, and values—as well as the city's economic, political, and institutional structures (Marcotullio et al. 2014). As a result, U.S. and Canadian

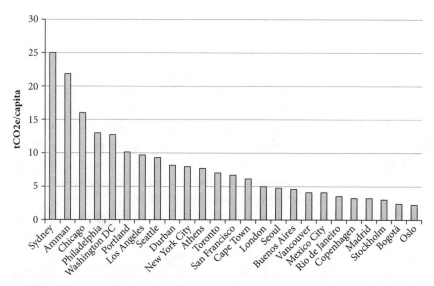

FIGURE 1.1 Range of GHG emission intensity in C40 cities, with New York City, Los Angeles, and Toronto highlighted. *Source:* C40 Greenhouse Gas Protocol for Cities Interactive Dashboard. Data from 2013. http://www.c40.org/research/open_data/5

cities are particularly egregious emitters (see figure 1.1). In the C40 Cities Climate Leadership Group, North American cities emit 60 percent more greenhouse gases than East Asian cities, and twice as much as European cities (ARUP 2014).

Climate change mitigation refers to actively taking the steps necessary to reduce GHG concentrations in the atmosphere in an effort to avoid the worst effects of climate change.[3] The negative effects of climate change and the cost of adapting to them will be much less severe if GHG emissions are greatly reduced: climate models estimate that very low levels of future emissions will lead to an average global temperature increase of 1.5°C, while very high levels of future emissions will lead to an average global temperature increase of 4°C (IPCC 2014b). The difference between these alternative futures is a function of dramatically reducing the level of energy consumption and use of fossil fuels (especially coal), increasing the use of renewable energy sources such as wind and solar, and making new technologies available (Van Vuuren et al. 2011). The difference between these two scenarios is also a dramatic and relatively rapid reduction in GHG emissions and subsequent concentration of CO_2 in the atmosphere: emissions must peak by 2020 and decline by 80–90 percent by the year 2050 to avoid the worst impacts of climate change.

Keeping global temperatures below a 2°C rise, which is what scientists say will avoid the most catastrophic effects of climate change, requires massive and rapid decarbonization. Recent research has detailed what this might look like: global CO_2 emissions have to be reduced by half every ten years from now until 2050, at which point they should have fallen from around 40 gigatons of CO_2 per year to 5 gigatons of CO_2 per year (Rockström et al. 2017). At the same time, emissions from land use like agriculture and deforestation have to fall to zero by 2050, and carbon capture technologies need to pull out 5 gigatons of CO_2 per year, which is about double the current sequestration rate of natural systems.

Clearly, swift, ambitious efforts to reduce GHG emissions are warranted and necessary to avoid the worst consequences of climate change. What is less clear is how to reorient the decision-making processes and incentives that underlie rapidly increasing emissions. This is a question the international community has been wrestling with, and in many ways ultimately failing to confront, for the past thirty years.

An Uncertain Global Response

In 1988 Jim Hansen, then a scientist with the NASA Goddard Institute for Space Studies, testified before the U.S. Congress about changes being detected in the Earth's atmosphere. Hansen told Congress unequivocally that systematic climate change was taking place, that human activities were the cause, and that the consequences were likely to be severe (Shabecoff 1988). His testimony served to initiate a policy discussion in the U.S. and, as evidence of climate change mounted, world leaders began to mobilize around the goal of creating a multilateral treaty able to limit global GHG emissions.

It seemed clear at the time that a global response to climate change was both an effective and feasible strategy for addressing the problem. Climate change is a global problem: its causes and consequences cross borders and boundaries, and no single country can secure large-scale reductions of GHG emissions on their own. The global community was also on a winning streak in environmental treaty making, having recently developed and implemented the highly successful Montreal Protocol on Substances That Deplete the Ozone Layer, in 1987. This success helped to solidify an understanding of the atmosphere as a global commons, and of "mega-multilateralism" (Hoffmann 2011) as the appropriate policy response. In keeping with this view, the world's initial response to climate change centered on compelling national governments to individually commit to reduce domestic GHG emissions to ensure the shared, planetary benefits of a more stable climate.

The first step in the attempt to craft a global response to climate change was the signing of the United Nations Framework Convention on Climate Change

(UNFCCC), which called for "stabilizing greenhouse gas concentrations in the atmosphere at a level that would prevent dangerous anthropogenic interference with the climate system" (UNFCCC 1992, 4). More than 150 countries, including both the U.S. and Canada, signed the UNFCCC at the Rio Earth Summit in 1992.

It was already clear to participants and observers that the United States was going to be a significant obstacle to internationally agreed-upon binding targets and timetables. Business and industry interests in the U.S. were skeptical that the costs of climate change mitigation were worth the benefits, and they were able to construct a bipartisan coalition in Washington that advocated for a cautious, voluntary approach to climate change mitigation (Hempel 2003). Rifts were also emerging between developed and developing countries. At the time, developed countries were producing two-thirds of global GHG emissions. Developing countries subsequently argued for a policy of "differentiated responsibilities" that would place a greater burden on developed countries to reduce GHG emissions and allow for fossil fuel–based industrialization and growth in developing countries. The result was a UNFCCC that lacked teeth and was largely aspirational: developed countries were asked to conduct inventories and pledge to voluntarily reduce their GHG emissions to 1990 levels by 2000.

Still, many remained optimistic that a more binding agreement could be struck. A series of negotiations followed the UNFCCC and culminated in the signing of the Kyoto Protocol in 1997. The Kyoto Protocol outlined a set of country-specific, binding GHG emission targets that included separate tracks for developing and developed countries. Developed countries (referred to as Annex I countries) committed, on average, to a 5 percent reduction in GHG emissions below 1990 levels by the 2008–2012 period; the U.S. and Canada committed at the time to 7 percent and 6 percent reductions, respectively. Developing countries (referred to as non-Annex I countries) were not given binding reduction targets. Rather, the agreement allowed for emissions trading between Annex I countries, and for Annex I countries to achieve their emission reduction goals by funding projects in non-Annex I countries via the Clean Development Mechanism. Importantly, if an Annex I country did not meet its targets, it would be required to either purchase the necessary credits from elsewhere or commit to additional reductions in the second commitment period (Leach 2011). To come into effect, the Kyoto Protocol required ratification by 55 percent of the parties to the UNFCCC and for those parties to collectively account for 55 percent of the developed countries' emissions.

The targets set in the Kyoto Protocol are very modest relative to the scale of reductions necessary to avoid the worst consequences of climate change. Still, there were a number of significant implementation challenges. First among these was the inability of the United States to ratify the agreement domestically and the federal government's long-standing resistance to binding targets of any kind. Further, the differentiation between Annex I and non-Annex I countries allowed the Kyoto Protocol to gain broad international support but it was a source of increased resistance from the United States (Hempel 2003). In 1997 the U.S. Congress passed the Byrd–Hegel Resolution, preventing the United States from ratifying any climate agreement that does not include shared expectations for developing countries. In 2001, within months of taking office, George W. Bush officially withdrew the United States from the Kyoto Protocol, announcing that the U.S. had "no interest" in implementing the treaty (Borger 2001). In February 2005 the Kyoto Protocol came into force after being ratified by Russia, but without the participation of the United States.

Canada successfully ratified the Kyoto Protocol in 2002, but it struggled to make good on its commitments. Indeed, Canada and the U.S. are the two Annex I countries that did not meet the 2010 Kyoto target and that are not on track to meet the 2020 Kyoto target (The Climate Institute 2015). In 2008, Canada's emissions were more than 30 percent above their Kyoto pledge (FCM and ICLEI 2010). Recognizing they were unable to meet their targets, Canada formally withdrew from the Kyoto Protocol in December 2011. While Canada's contributions to global GHG emissions are a fraction of those from the U.S., they contributed to the stymied negotiation process and their withdrawal from Kyoto defused optimism for an effective global response to climate change. Further, normative issues surrounding who should pay for climate change mitigation, and whether some countries may have a right to continue to emit GHG emissions, continued to plague efforts to achieve binding global commitments. Large developing countries remained hesitant to take significant steps to reduce GHG emissions and "consistently signaled that they prioritize development goals over climate protection" (Hoffmann 2011, 15). Given their relatively small contribution to the cumulative levels of greenhouse gases in their atmosphere, they argue that they should not be asked to forego any of the benefits that come from the exploitation of cheap and available fossil fuels. With the rapid increase in GHG emissions from countries such as China, India, and Brazil, such arguments may be difficult to sustain.

The recently negotiated agreement at the 21st Conference of the Parties, held in Paris in December 2015, inserted cautious optimism for the international process. The U.S. position on climate change shifted under President Obama, and his administration helped to lead negotiations in Paris. Here, leaders committed in principle to limiting climate change to a global temperature increase of 2°C. Achieving this goal requires that global GHG emissions peak "as soon as possible" and rapidly approach zero by 2050. The Obama administration committed to reducing GHG emissions in the U.S. by 17 percent below 2005 levels by 2020 and introduced stricter fuel efficiency standards for new vehicles, new emission regulations for power plants, and investments in renewable energy sources. The Paris Agreement does not include binding country-level emission reduction targets, in part as a way of avoiding the need for ratification by the U.S. Congress and soothing tensions between more and less developed countries. Rather, signatories commit to monitor GHG emissions and developing plans for reducing domestic emissions. The Paris Agreement therefore shifted focus away from a collective emissions reduction target to individually determined voluntary pledges and shared review processes.

While it created some optimism, the Paris Agreement also comes with a significant amount of uncertainty around implementation and operationalization. Participants agreed to a rather vague temperature target with a loosely designed set of reporting requirements for voluntary and self-determined reduction pledges. The mechanisms for implementation and accountability are therefore highly decentralized and devolved, requiring that individual countries define and pursue an emissions reduction target independently. Most damningly, the Trump administration announced in June 2017 its intention to withdraw the U.S. from the Paris Agreement. Indeed, President Donald Trump has pledged to reinvest in coal and coal-related technologies and roll back many of the Obama administration's initiatives to reduce GHG emissions. Canada has gone through the reverse transformation, with Prime Minister Trudeau announcing in 2016 a national GHG emissions cap and trade policy, and articulating strong support for the Paris Climate Agreement. Without participation from the U.S.—the largest GHG emitter and an important leader on the global stage—prospects for significant and tangible global action on climate change are dim.

As a result, there is a growing sense that the mega-multilateral approach to climate change mitigation may be coming to the end of its useful life. While the Paris Climate Agreement maintains a multilateral approach to climate change, there is no longer a globally agreed-upon, defined collective response with country-level responsibilities. Further, the multilateral approach has failed to produce any meaningful reductions in GHG emissions: global emissions have actually increased by 47 percent between 1990 and 2012 (WRI, CAIT 2014). The process has also failed to resolve many of the long-standing challenges to climate change mitigation: the need for very flexible and adaptive policy instruments, a clear link between costs and benefits, and the need for fundamental shifts in current lifestyles. As Hoffmann observes,

> the international community has struggled mostly in vain to overcome the obstacles to collective action and devise an effective multilateral response to climate change. . . . It may be time to concede that there is a mismatch between this type of treaty-making and the problem of cli-

mate change; that global treaty-making, as attempted in the last two decades, cannot catalyze the societal and economic transformation necessary to avoid the potentially catastrophic consequences of climate change. (2011, 12)

Fortunately, as nations have struggled to forge a coordinated path forward to address climate change, and as the current approach is surrounded by uncertainty and decentralization, there has been a groundswell of climate change mitigation ambition in cities around the world. In many countries—including Canada and the U.S., but extending to Mexico, India, and China—cities have been the first to develop climate change mitigation plans and pledges, absent any requirements from regional or national governments. What was once considered an inherently global challenge requiring global solutions is increasingly in the hands of city governments.

The Emergence of an Urban Response

In the face of policy stagnation and uncertainty at the international and national levels, city governments have emerged as climate change leaders. As Paris mayor Anne Hidalgo declared in 2015, "cities are not waiting for [national leaders] to give us the solutions. We are moving ahead and making a solution possible" (Scruggs 2015). Indeed, thousands of city governments around the world have committed to reducing their GHG emissions; some have even pledged to be "carbon neutral," meaning they will make no net contribution to global CO_2 levels. Cities are now active participants in global climate change policy: their ambition and enthusiasm have become a source of hope and optimism in an often-dismal climate change policy landscape. City governments are now seen as legitimate actors in the climate change arena, and a growing cohort of city leaders view climate change as an appropriate undertaking for city governments.

Cities in the U.S. and Canada have been at the forefront of the urban response to climate change. Toronto hosted the first international meeting on climate change—the World Conference on the Changing Atmosphere—in 1988, and became the very first jurisdiction to develop a GHG reduction target and plan. Toronto and Vancouver are both founding members of the C40 Cities Climate Leadership Group, an international network of large cities committed to acting on climate change. David Miller, former mayor of Toronto, served as the second president of the C40 from 2008–2010. Eleven of the largest cities in the U.S. are members of the C40 Cities Climate Leadership Group, including New York City, Los Angeles, Boston, Chicago, San Francisco, and Seattle. Michael Bloomberg, a former mayor of New York City, chaired the C40 from 2010–2013 and is now the UN Special Envoy for Climate Action.

Beyond the C40, there are more than three hundred Canadian municipalities in the Federation of Canadian Municipalities's Partners for Climate Protection program, representing more than 65 percent of the Canadian population (Federation of Canadian Municipalities 2017). Nearly half of ICLEI's Green Climate Cities (GCC) program membership comes from the U.S.; here, member cities represent 25 percent of the U.S. population (75 million people) and forty-seven states (ICLEI 2015). There are also efforts to foster coordination and collaboration on climate change mitigation between U.S. and Canadian cities via The Cities Initiative, a network of 120 mayors founded in 2003 by Mayor Richard Daley of Chicago. This group has committed to collectively reducing 30 million tonnes of equivalent CO_2 by 2020.

The ambition of city governments to tackle climate change mitigation is clear, but the evolution of this commitment reveals an unfolding agenda. Since the actions of a few pioneering cities in the early 1990s, the urban response to climate change has broadened in scope and sophistication.

When climate change came on to the public agenda in the early 1990s, several city governments were quick to take independent steps to set GHG emission reduction goals ahead of any global or national targets. The 1988 World Conference on the Changing Atmosphere inspired these early actions. Over three hundred scientists and policymakers from around the world met in Toronto to discuss the implications of, and solutions for, the emerging problem of climate change. Following the conference, the City of Toronto passed a unanimous resolution to reduce citywide CO_2 emissions 20 percent below 1988 levels by 2005 (Harvey 1993). The City went on to establish an Energy Efficiency Office and began to outline a set of options and strategies for achieving its reduction goals (City of Toronto 1991). The City also conducted a comprehensive GHG emissions inventory and outlined emission reduction programs such as community-wide energy efficiency retrofits and district heating and cooling systems. Portland, Oregon, was similarly motivated after city representatives attended the meeting in Toronto. They adopted a carbon-reduction plan in 1993 that aimed to reduce emissions 20 percent below 1988 levels by 2010. Their plan included a detailed set of implementation strategies, outlining a focus on transportation, energy efficiency, renewable energy resources, recycling, and tree planting (Boswell, Greve, and Seale 2012).

These early actions planted the seeds for the Urban CO_2 Project, led by the International Council on Local Environmental Initiatives (ICLEI). The Urban CO_2 Project emerged from the work of twelve large cities in Europe and North America as a first effort to encourage learning and the sharing of best practices between cities (Harvey 1993). The twelve cities developed targets and strategies independently but were given opportunities to share them with one another and to access ICLEI's expertise and technical capacities. Members were expected to submit by 1993 an action plan for reducing their emissions and to commit to reducing CO_2 emissions from energy use within their jurisdiction by 10–20 percent (Lambright, Chjangnon, and Harvey 1996).

The Urban CO_2 Project was a success. By 1996 it had been renamed Cities for Climate Protection (CCP, what would later be renamed the Green Climate Cities program) and had one hundred participating cities. By the end of the decade more than four hundred cities were CCP members, 109 of which were in the United States (U.S. Environmental Protection Agency 2001). At this point the membership expectations were more explicit: in joining the CCP, city governments were expected to pass a resolution through their city councils committing to climate change mitigation and a GHG reduction target. In Canada, efforts to establish city networks for climate change mitigation were initiated in 1994 by ICLEI and the Federation of Canadian Municipalities: ICLEI launched the Partners for Climate Protection (PCP), while the Federation of Canadian Municipalities established the "20% Club," with six founding cities committed to reducing city government emissions 20 percent below 1994 levels. The two programs merged in 1998. Intercity networks also began to form in Europe during this time, such as the Climate Alliance, CCP-Europe, and Energe Cités (Bulkeley 2013, 76).

During this first phase of the urban response to climate change, city governments quickly emerged as willing to act on climate change without being compelled to do so by higher levels of government. There was eagerness among city governments and supporters to harness that ambition, learn from one another, and create opportunities for networking and support. While the ambitions of city governments at this time were often community-wide, much of the early action focused on reducing emissions from city government operations (Betsill and Bulkeley 2007). Many of the emission reduction targets cities set mirrored those being developed at the international level, and so were relatively modest.

By 2001 the international climate change policy process faced extreme uncertainty and pessimism. The United States had withdrawn from the Kyoto Protocol, and newly industrializing countries like India and China were hesitant participants at best. The landscape of local climate change action was changing rapidly: the number of cities taking action quickly grew, more intercity networks began to emerge, and the urban climate change policy agenda expanded and solidified.

During this time participation in ICLEI's GCC program grew by leaps and bounds: by 2006 there were 546 local governments involved, representing 243 million people in 27 countries. As early as 2002 more than one-third of GCC members were from outside Europe and North America, including 45 cities in Asia,

Africa, and South America (Yienger, Brown, and Skinner 2002). A similar process was unfolding in Canada. By 2007 the Partners for Climate Protection program had recruited 155 member cities representing 70 percent of Canada's population. European networks had almost 1,400 European cities and towns by the early 2000s (Kern and Bulkeley 2009). The diversity of cities participating also increased rapidly, from small midwestern U.S. cities like Urbana, Illinois, to sprawling Sunbelt U.S. cities like Tallahassee, Florida, to rapidly urbanizing African cities like Dar es Salaam, Tanzania.

New intercity networks formed, the most prominent of which was the C40 Cities Climate Leadership Group. In 2005, then-mayor of London Ken Livingstone invited leaders from eighteen large cities to convene as The Climate Group as a parallel event to the G8 Climate Change Roundtable being held in Scotland. In 2007, this network entered into a partnership with the Clinton Climate Initiative (CCI) and expanded its membership to include forty of the largest cities in the world. The C40 differed from ICLEI's networks in that it targeted large "megacities" and had a more explicitly political purpose. A central focus was commitment from members to respond to climate change and reduce emissions, but the C40 also sought to act as a medium for demonstrating the power and legitimacy of cities to respond to climate change. For example, they encourage networking among member cities, and have regular meetings for members.

This emphasis by the C40 on the political dimensions of city engagement with climate change is reflective of a broader trend during this time toward what Betsill and Bulkeley (2007) refer to as a shift toward "strategic urbanism." While initially city governments were likely to focus policy action on their own emissions, they subsequently began to focus more attention on reducing community-wide emissions. It wasn't long before cities realized that efforts to reduce community-wide GHG emissions were often in tension with other political priorities (Bulkeley and Betsill 2003) and that new strategies and partnerships were needed in order to be effective (Hodson and Marvin 2010; Castán Broto and Bulkeley 2013).

In 2005 the Kyoto Protocol came into force, paradoxically spurring a flurry of action by mayors rather than national governments. When it was clear that the United States would not be ratifying the treaty, the then-mayor of Seattle, Greg Nickels, issued a challenge to U.S. mayors to reduce GHG emissions in their cities. By 2011 this call had attracted over 180 U.S. mayors and led to the signing of the Climate Protection Agreement (C. D. Gore and Robinson 2009). Also in 2005, a group of twelve mayors from large cities around the world formed the World Mayors Council on Climate Change, which grew to include seventy cities by 2012 (World Mayors Council n.d.). In 2009, the European Covenant of Mayors emerged, which requires signatories to go beyond the EU GHG emission reduction targets, and by 2011 had more than 2,000 members (Bulkeley and Betsill

2013). Like the C40, these networks have typically sought to call attention to the power and potential of city governments to respond to climate change and encourage greater action from higher levels of government.

This wave of network activity also led to local policy action. For example, in 2007 the mayors of New York City, Toronto, and Los Angeles each released their plans for reducing GHG emissions. All three focused heavily on strategies for low-carbon energy generation, energy efficiency measures for the built environment, and investments in active and public transportation. Toronto took as its guide the targets being promoted at the international level, moving from the 6 percent reduction below 1990 levels by 2012 Kyoto target to a goal of an 80 percent reduction by 2050.[4] Los Angeles also used 1990 as its baseline year but set a goal of reducing emissions by 35 percent by 2030. New York City chose its peak GHG emissions in 2005 as its baseline, therefore perhaps making the most ambitious goal of reducing 30 percent by 2030.

Importantly, plans like these also signal a shift toward more strategic, executive-led planning for climate change mitigation. Where previously such efforts may have been the purview of a technical agency in the city, such as the Energy and Environment Division in the case of Toronto, we now see mayors and their executive teams begin to assume a greater leadership role in setting a vision for their city that includes an ambitious GHG emissions reduction target. The C40 also raised the profile of mayors as important climate change champions, providing them with an international platform from which to demonstrate their commitment to addressing climate change, highlight the actions their cities are taking, and access technical support and resources.

The number of cities engaging with climate change continues to grow, such that today several thousand city governments, representing all parts of the globe, are engaged in some way with climate change mitigation.[5] Some policy attention is also beginning to shift to climate change adaptation, through forums such as the Durban Adaptation Charter for Local Governments of 2011 and the European Mayors Adapt Initiative of 2014. There has also been a distinctive shift in the political position of city governments at the international level and the consolidation of targets and protocols around shared sets of practices.

City governments have become legitimate climate change actors in the international arena and are taking steps to leverage this legitimacy to influence policy at multiple levels. Two important events took place in 2010 that illustrate this shift. The World Mayors Summit on Climate Change was held on November 21, 2010, in the lead-up to COP16 in Cancún, Mexico. One outcome of this meeting was the Mexico City Pact, an agreement among 138 world mayors to commit to act on climate change, including by reducing GHG emissions, and a call to international negotiators to recognize that cities are strategic in combatting climate change. The commitments would reduce GHG emissions by 740 million tons annually, more than Mexico's annual emissions. The effort was an attempt to define the role of cities more centrally and explicitly.

At the subsequent talks in Cancún, local governments were recognized for the first time as governmental stakeholders in the negotiations. Christiana Figueres, executive secretary of the UNFCCC, stated: "legislators, mayors and city leaders are at the forefront of the efforts to care for the planet, implementing both mitigation and adaptation efforts in the cities" (World Mayors Council on Climate Change 2010a). This shift caught the attention of UN Secretary Ban Ki-moon, who noted that in future negotiations cities would be recognized formally, and this has been the case. At a lead-up meeting to the 2015 COP21 in Paris, city leaders were told by UN Assistant Secretary-General Janos Pasztor that they are "on the frontlines of the climate challenge" and that national governments "need your support to raise ambitions" (United Nations 2015). Indeed, more than a thousand city leaders attended the Climate Summit for Local Leaders held at COP21. Also in 2010, the then-mayor of New York City, Michael Bloomberg, became chair of the C40. Bloomberg was able to shift and consolidate the focus of the C40 to "converge around a shared understanding of climate change as an urban, rather than a market, problem. Climate change, in other words, is not only amenable to but actually *requires* an urban (read city-led) response" (Gordon 2015). This shift helped to further bolster the legitimacy of cities in the international process, including Bloomberg's recent appointment by Secretary Ban Ki-moon as Special Envoy for Cities and Climate Change.

As city governments have become more engaged with climate change mitigation, it has become increasingly clear to national governments that cities will be necessary partners for achieving any national GHG emissions reduction target as well. The ambition of city governments helped to propel the climate change negotiations between the U.S. and China. The lead climate change negotiators for the United States and China highlighted the role cities play in implementing federal climate change programs and the momentum created by their independent actions and commitments for a historic agreement between the two countries (Davenport 2015).

City governments have become more confident in their role in climate change mitigation and are more explicitly leveraging their authority to influence or even replace national policy. For example, in 2014 Eric Garcetti, mayor of Los Angeles, helped to launch the Mayors National Climate Action Agenda, a coalition of twenty-seven U.S. mayors committed to reducing urban GHG emissions and advocating for greater action from the federal government. In response to local demands, the U.S. government developed the Energy Efficiency and Conservation Block Grant program in 2009 as part of the American Recovery and Reinvestment Act. The program funds energy efficiency programs and requires that such programs create local jobs. When President Trump committed to withdrawing the United States from the Paris Climate Agreement, more than 160 U.S. mayors and 4,700 mayors globally pledged to remain committed to the goals (Halper 2017). Canadian prime minister Justin Trudeau expanded federal government funding to the PCP program following COP21, committing C$75 million to support the efforts of cities.

In addition to the shift in the political position of city governments in climate change policy, a more consolidated set of practices and protocols is now available to them for planning and program design. There is a particular emphasis on data collection and transparency in many city networks, and city governments and their supporters have invested significant effort in developing tools for measuring and reporting city-level GHG emissions and documenting the steps city governments are taking to reduce emissions. The 2010 World Mayors Summit established the first shared registry for city governments to report their GHG emissions and climate change policies and programs, the carbonn Cities Climate Registry (cCCR). As chair of the C40, Bloomberg brought with him a passion for data and, in 2014, the C40 helped to lead the development of a Global Compact of Mayors at the COP20 in Lima. The compact committed cities to using shared reporting and monitoring platforms and to adopting a target of reducing GHG emissions 80 percent by 2050, mirroring the most ambitious of the international targets. There are currently 640 cities that have signed the Compact, 23 of which are from Canada (including Toronto) and 133 of which are from the United States (including New York City and Los Angeles).

In 2016, a new "mega-network" of world mayors, the Global Covenant of Mayors for Climate & Energy, consolidated existing agreements and networks (including the Global Compact of Mayors) and provides its 7,000 members with a shared reporting and monitoring platform. This group also developed a common protocol for urban GHG emission inventories, the Global Protocol for Community-Scale GHG Emissions Inventories (GPC). While the convention has been that cities focus on monitoring and reporting Scope 1 and 2 emissions, the GPC suggests that cities also report Scope 3 emissions from waste (Fong et al. 2015). Many cities have already adopted the GPC for their inventories and reporting, including New York City, Los Angeles, and Toronto. Such data are not always collected by city governments, or may be collected in different ways in different cities, complicating efforts to develop a clear picture of the GHG emissions contribution of cities.

Climate change mitigation policy can no longer be seen as the sole territory of the mega-multilateral negotiation process; nor is it the pet project of a handful of pioneering cities. City governments have moved from acknowledged contributors

TABLE 1.1 The scaling up and out of the urban climate change response, 1990–2016

TIMEFRAME	SCALE OF CITY INVOLVEMENT	POLICY AGENDA	EXAMPLES
1990–2000	From 0 to more than 400 cities Largely in developed countries Dominated by large cities	Target setting by individual cities Early networking between cities	Toronto's GHG emission reduction target and plan Development of ICLEI's Cities for Climate Protection Program
2000–2010	More than 1,000 cities More involvement from developing countries Large and small cities	City numbers and networks multiply and expand More strategic, often mayor-led, climate action plans by large cities	C40 Cities Climate Leadership Group PlaNYC in New York City
2010–2016	Thousands of cities Cities from around the world Large and small cities	Greater international influence Advocating for national action Global standardization of targets and protocols	Cities recognized as stakeholder at COP15 in Copenhagen EECBG in the U.S. Global Protocol for Community-Scale GHG Emissions Inventories

of climate change solutions to legitimate actors in the international arena, armed with increasingly sophisticated tools and resources (table 1.1). As Hoffmann notes, "the center of gravity in the global response to climate change is shifting" (2011, p. 5), and thousands of ambitious city governments around the world have emerged as an attractive force. The actions cities are taking to respond to climate change stand in stark contrast to the halting progress at the international level and have placed them on the world stage (Lutsey and Sperling 2008; Hjerpe and Nasiritousi 2015). This shifting center of gravity is also reflected at the national level in the U.S. and Canada. City governments have come to think differently about their role in climate change mitigation and of their responsibility to address a global problem.

Why Cities Are Responding

The ambitions and leadership of city governments on climate change mitigation are exciting—they have reinvigorated the climate change community and have led many to speculate that cities will be the solution to climate change. As we have learned more about the global drivers of GHG emissions, it has also become clear that cities *must* be central to any serious mitigation efforts. Cities are responsible for a significant portion of global GHG emissions due to the concentration of social and economic activity in their borders, and they have significant authority over many of the processes, behaviors, and investments that drive GHG emissions.

Still, in many ways a central role for cities in climate change mitigation is unintuitive: climate change is a global problem that extends beyond the border of any single city. City governments have a clear incentive to free ride on the progress other cities make in reducing their GHG emissions, and in most cases they are under no obligation to take up the issue. It is also not immediately clear why a policy issue that is so intractable at the federal level would find a foothold at the local level, where conventional wisdom predicts an overriding emphasis on growth and economic development (Molotch 1976; Peterson 1981). While cities are relatively familiar with sustainability initiatives through programs such as Local Agenda 21, North American cities have typically lagged behind in such efforts (Portney 2013, 23). Further, compared to the broader sustainability agenda, reducing GHG emissions represents a task for city governments that primarily targets energy use, energy supplies, and transportation systems.

The dissonance between our expectations of cities and their climate change ambitions has led to a significant body of work that illuminates the myriad reasons cities have for tackling climate change. These include opportunities for leadership, entrepreneurship, and city branding; a sense of responsibility to address a threat; a desire to capture the local co-benefits from reducing GHG emissions; and responding to either policy action or inaction by other levels of government (Hughes 2017). These reasons are, of course, not mutually exclusive or present in every city, but they are the types of political and economic forces underlying local action on climate change. While understanding the motivations for cities to address climate change provides political context for their work, it does not ultimately offer insight into how city governments might go from their ambitions and commitments to realized change in their cities.

Opportunities for Leadership

In a recent survey of 350 ICLEI GCC members from around the world, the top reason given by city governments for responding to climate change was to "demonstrate leadership globally, nationally, or regionally" (Aylett 2014, 22). Indeed, climate change has become something of a celebrity issue for city mayors, and bold action has landed several mayors in the national or even international spotlight. One interesting example is the city of Grand Rapids, Michigan. A small city of less than 200,000 people, its commitment to climate change gave former mayor George Heartwell international exposure. The World Wildlife Fund sponsored his participation in the 2015 Paris climate talks, where he represented the Compact of Mayors (Vande Bunte 2015). Another prominent example is former mayor of New York City, Michael Bloomberg. Following New York City's successes with climate change policy and planning, Bloomberg has taken up a position at the United Nations as a global spokesperson for cities and climate change.

Mayors with larger ambitions for political office or greater exposure might be able to raise their profile by taking innovative action to address climate change, demonstrating their leadership abilities and forward-thinking vision. For example, some have speculated that former mayor of Los Angeles Antonio Villaraigosa's interest in environmental issues was driven in part by a desire to run for governor of California, a position increasingly defined by environmental leadership. Indeed, Villaraigosa ran for governor in 2018, but lost in the Democratic primary. Climate change has relevance in state, provincial, and national political settings that other urban policy areas, like service delivery or even crime, may not carry. These dynamics are particularly evident in the rise of global networks of mayors committed to addressing climate change such as the Mayor's Compact and the C40 network. Participation and engagement in these global networks can subsequently expose mayors to new norms, new ideas, and new sets of technical resources that can shape the decisions city governments make about their own approach to reducing GHG emissions and the outcomes they are likely to achieve as a result.

Beyond individual ambitions, cities want to be seen as leaders themselves and are often competing with one another to be at the cutting edge. Cities reference a desire to be the "greenest city in the world" (Vancouver, Canada), the "first carbon neutral city in the world" (Adelaide, Australia; Copenhagen, Denmark), the "most sustainable port city in the world" (Rotterdam, The Netherlands), or the "first sustainable 21st century city" (New York City). Cities track who has the most LEED-certified "green" buildings or the highest levels of transit ridership. Leading on climate change can be a way to demonstrate a city's ambition, creativity, political strength, and innovation.

Entrepreneurial Cities and City Branding

As engines of growth and innovation, cities also translate, and are shaped by, global economic and political forces (Clarke and Gaile 1998; Sassen 2000). The global embeddedness of cities (particularly large cities such as Toronto, Los Angeles, and New York City), and the competition among them for status, reputation, and resources, propels city governments to be increasingly entrepreneurial in their governing strategies.

Climate change leadership can provide a city with a competitive edge in attracting residents and investment. Where environmental policies were once thought of as a liability to cities operating in highly competitive environments, there is growing evidence that such "extra-economic" factors (i.e., being a "sustainable city") are helping to brand cities and urban spaces in ways that are advantageous in a global marketplace (Jessop and Sum 2000; Florida 2002). Responding to climate change provides leadership and entrepreneurial opportunities to cities looking to craft an image of sustainability and competitiveness (Blakeley and Green Leigh 2010; Koven and Lyons 2010). Indeed, their efforts on climate change can be seen as part of a broader phenomenon in which city governments are taking an entrepreneurial role on issues such as regional transportation, trade, disaster response, immigration, and poverty (Clarke and Gaile 1998; Kettl 2000; B. R. Barber 2013; Beal and Pinson 2013). For example, the Global Covenant of Mayors states that its efforts to make the climate change work of cities more transparent and professionalized will provide investors with "the ability to see the actions cities have been taking are having lasting, verifiable, and most importantly, *investable* impact" (Global Covenant of Mayors n.d.).

Copenhagen has seen exactly these benefits since deciding to be a leader on climate change, pledging in 2012 to be the world's first carbon neutral capital city by 2025, as well as the world's best bicycling city, and making progress toward accomplishing it. The city has made broad changes to its energy and transportation systems, adding a district cooling network and nearly doubling its bicycling infrastructure (Gössling 2013). These changes haven't slowed city growth; if anything they have attracted residents. In 2016 the city experienced record population growth, adding 4,300 new housing units and adding around 10,00 new residents each year (W. 2017). Residents also find the changes exciting: Copenhagen Lord Mayor Frank Jensen is quoted as saying, "Copenhageners like the ambition, they like being part of the idea of going green for the whole city" (Gerdes 2013). Responding to climate change can help a city craft an attractive "brand" that generates investment and attracts residents.

Capturing Local Co-Benefits

Responding to climate change can be attractive to city governments because of the multiple local benefits it provides in addition to reducing GHG emissions. Programs and investments that reduce GHG emissions often have other benefits that may be of equal or greater importance to a city and its residents, such as reducing air pollution, reducing costs, managing growth, reducing congestion, and increasing mobility. For example, investing in public transportation not only reduces GHG emissions but also improves congestion and air quality and increases mobility for urban residents. Investing in energy efficiency measures saves money by reducing electricity bills and avoiding the need to expand generation capacity.

New programs and services in support of climate change mitigation can create new jobs in a city. For example, a study in Los Angeles found that every $1 million spent on energy efficiency creates sixteen job-years for the city compared to residential construction, which creates 10.7 job-years per $1 million (DeShazo, Turek, and Samulon 2014).

The attractiveness of local co-benefits is driven in part by the receding interest of national governments in many places in both urban and climate change policy and the subsequent venue shopping this has prompted. International and national organizations such as the C40, ICLEI, and the NRDC actively campaign to encourage climate action in cities in the wake of federal policy stagnation, and cities are likely to see this as an opportunity for investment in an increasingly resource-constrained environment.

City governments have been savvy in identifying the policy and economic co-benefits of addressing climate change, often tailoring their climate change plans to produce benefits beyond GHG reductions (Betsill 2001). Indeed, promoting sustainable urban development more generally and improving community quality of life were key motivations for 57 percent and 53 percent of the ICLEI GCC survey respondents, respectively (Aylett 2014, 22). Framing climate change policy in a way that highlights its multiple benefits can also be politically useful, as it expands the constituency likely to support and mobilize around local action. For example, vulnerable populations such as children, the elderly, the poor, and racial minorities may stand to benefit the most from the co-benefits of climate change mitigation (Younger et al. 2008).

A desire to capture the local co-benefits of climate change mitigation is clearly evident in many urban climate change plans. Cities often emphasize the benefits for residents' health, the local economy, and livability (e.g., equity, congestion, green space), highlighting the opportunities that climate change might represent rather than simply its consequences. Los Angeles sums up this motivation nicely in its climate change plan *Green LA*, writing:

> While the risks associated with climate change are high, the benefits of acting today are largely positive. Reductions in carbon emissions will improve air quality, create a more livable city, and invent cutting-edge green technology that can be marketed to the global community. Viewed properly, the threat of climate change is really an opportunity to transform Los Angeles into the greenest big city in America—a model of urban sustainability for the 21st century. (City of Los Angeles 2007, 3)

A lack of valuable co-benefits can also serve to dissuade city governments from responding to climate change. For example, cities with a manufacturing base or other source of carbon-intensive activities are less likely to have a climate change plan in place (Zahran et al. 2008; Krause 2011a, 2011b, 2012; Sharp, Daley, and Lynch 2011). If there isn't something in it for cities, they are not likely to take action to address climate change.

Responsibility to Address a Threat

In many ways, city governments are on the front lines of climate change. The high land values, dense populations, and concentration of economic activity in cities mean they are highly vulnerable to the impacts of climate change such as sea level rise, heavy precipitation, drought, and extreme hot and cold temperatures. Cities are often where the human and financial costs of these events are felt most acutely. The proximity of cities to climate change exposures and vulnerability can compel city governments to take action, perhaps in a way that is distinct from regional and national governments. As described earlier, cities have gained increasing confidence in their ability and responsibility to respond to climate change. Faced with failed national and international policy efforts, they are willing to invest municipal resources into the collective effort to reduce GHG emissions. City governments are increasingly aware of the contribution of cities to overall GHG emissions and view themselves as having a role to play in solving the collective problem.

We see evidence of this motivation in much of the discourse surrounding the urban response to climate change. At the "What Works Cities" summit held in March 2017, Michael Bloomberg referenced the front line position of cities, declaring "cities are where the people are, where the problems are, and where the solutions are" (Sunlight Foundation 2017). Many cities have recently experienced severe weather events, which help make climate change a more tangible problem for both decision makers and the public. For example, Hurricane Sandy in New York City and a major ice storm in Toronto both led to renewed attention to climate change in the two cities. While there is evidence in cities' plans and rhetoric that city governments are aware of the threat climate change poses to their city, and theoretical reasons we would expect to see this awareness lead to action, there is not a systematic link between a cities' vulnerability to climate change and their likelihood of responding (Zahran et al. 2008; Krause 2012).

This sense of responsibility to address climate change is more likely driven by public opinion and support, and demand for action on climate change is on the rise: recent surveys in both Canada and the U.S. show that a majority of the population is concerned about climate change and support measures to reduce GHG emissions (Borick, Lachapelle, and Rabe 2011; Environics Institute for Survey Research 2015; Leiserowitz et al. 2016). With local governments often considered more accessible than regional or federal governments, this public support for climate change may be more likely to motivate action by city governments.

For example, decision makers from both New York City and Toronto view the release in 2007 of Al Gore's film *An Inconvenient Truth* as an important milestone for climate change action in their city,[6] though there was little in the way of federal policy action that followed in either case.

Studies have shown that the cities that are more likely to house and attract constituents interested in environmental issues are often the cities taking action. For example, in the U.S., cities that are better educated, wealthier, and home to active environmental groups are on the whole more likely to have a climate change plan in place (Brody et al. 2008; Zahran et al. 2008; Portney and Berry 2010; Krause 2011a, 2011b, 2012; Sharp, Daley, and Lynch 2011). Institutional mechanisms for interest group and community engagement in city decision making also make the adoption of climate change plans more likely (Sharp, Daley, and Lynch 2011). The city's sense of responsibility to address the threat of climate change may be a direct response to local constituent and interest group concerns.

Responding to (In)Action at Other Levels of Government

City governments may also be compelled to reduce their GHG emissions because of incentives or requirements coming from other levels of government. While national governments have been reluctant to commit to international agreements, some are beginning to develop domestic measures to reduce energy use or curb emissions. In many cases there has been a trend to devolve responsibility for climate change to the local level more or less explicitly (Mazmanian and Kraft 2009; Schreurs 2010; Cohen and Miller 2012). For example, in the U.S., President Obama capped the quantity of GHG emissions power plants are allowed to produce, many of which are owned and operated by municipal governments. Nearly half of state governments in the U.S. have a renewable portfolio standard for energy supplies, and many have energy efficiency codes—both of which have implications for city governments. While these programs are under attack from the Trump administration, it is clear that cities are implicated in many federal efforts to curb GHG emissions.

Federal and subnational governments have been slow to act in the U.S. and Canada. While city governments might take cues from higher levels of government, they are very rarely responding directly to a set of requirements for climate change mitigation. There is mixed evidence as to whether state and provincial policies play a significant role in shaping local policymaking and priorities. Some studies have shown that the adoption of climate change policies by city governments is more likely when higher levels of government have standards and goals of their own (Schreurs 2010; Krause 2012; Homsy and Warner 2015). There is also evidence that city-level factors are more important in determining how and whether cities develop policies to reduce GHG emissions (Lee and Koski 2012; E. Schwartz 2016).

Cities are also influenced by changes in the international climate change policy landscape, such as the outcomes of international meetings or reports from the Intergovernmental Panel on Climate Change. One clear example of this is former mayor Greg Nickels's call to action following the failure of the U.S. to ratify the Kyoto treaty. Intercity networks provide opportunities for cities to gain technical expertise and learn from other cities' experiences, but evidence for whether such networks have a tangible influence on local decision making is mixed at best (Bulkeley and Betsill 2003; Krause 2012; Lee 2012).

Conversely, cities may take a leadership role in addressing climate change with the hope of motivating other levels of government. City governments see their own efforts to reduce GHG emissions as a way to pressure or inspire regional or federal governments to do the same.[7] Indeed, this is the explicit aim of many international city climate change networks. City governments are keenly aware that support and assistance from higher levels of government are necessary for them to meet their goals, and a bold commitment to meet the climate change challenge raises awareness and provides leverage for those who would like to see action at higher levels. This strategy can be effective. Cities are important laboratories for experimenting with new ways of doing things and can demonstrate the feasibility of novel approaches (Shipan and Volden 2006; Bulkeley and Castán Broto 2013). Los Angeles and Toronto have both adopted green building codes that require enhanced energy efficiency measures that were subsequently adopted by their respective state or provincial government. New York City hopes that its stated desire for renewable energy supplies will influence the choices the state government and private utilities make about future energy portfolios. If cities can find ways to innovate and demonstrate new approaches, others may be more willing to follow suit.

The Emerging Gap between Urban Ambition and Progress

Given the challenges inherent in climate change mitigation, it is perhaps unsurprising that city governments have a mixed record on reducing GHG emissions. Networks like ICLEI and the C40 are eager to demonstrate the progress their member cities are making, touting the thousands of programs and policies member cities have developed in an effort to reduce GHG emissions. For example, ICLEI Canada's PCP program reports that its members "have undertaken more than 800 GHG-reduction projects that represent more than C$2.3 billion in investment and 1.8 million tonnes in annual GHG reductions" (Federation of Canadian Municipalities 2015b, 4). Likewise, the scale and scope of action in C40 cities is clearly increasing, as is their use of monitoring and reporting tools. The C40 reports that member cities have over 9,000 different actions in place to respond to climate change, with at least half of these targeting the city as a whole (ARUP 2015). Energy conservation is an area of particular success for cities. A 2010 survey of local governments in the U.S. found that 81 percent are actively working to reduce energy use in buildings (Svara, Read, and Moulder 2011).

The progress cities have made can also be seen as evidence of a glass half empty. Taking concrete action to reduce GHG emissions has been piecemeal and slow for many cities. In Canada, only 41 percent of PCP members have a GHG emissions reduction target, 34 percent have an action plan, and just 12 percent report implementing their action plan. Looking at the figures from ICLEI's GCC program, nearly half of GCC members have not taken any steps toward their climate change mitigation goals, and less than 10 percent have completed all five of the milestones ICLEI lays out for GCC member cities (Reams, Clinton, and Lam 2012). In fact, a recent study found that GCC membership declined 22 percent between 2010 and 2012 and a number of cities had abandoned their climate protection efforts altogether (Krause, Yi, and Feiock 2015). While the overall trajectory is one of greater urban engagement with climate change policy, there is variation and ongoing evaluation of these efforts that may be tied to political cycles or administrative restructuring.

A similar story can be told about the C40. While the number of actions cities are taking, and the proportion of city governments meeting their membership obligations, continues to increase at a promising rate, only half have gone through the initial steps of developing a plan, setting a target, and completing an inventory (Gordon 2015). Further, while the great majority of urban emissions come from activities outside of city government operations, only half of the steps C40 cities are taking to reduce GHG emissions tackle citywide processes and systems (ARUP 2015).

Looking beyond formal city networks also reveals a gap between ambition and progress. In a survey of city governments in Indiana, Krause (2011b) found that while 20 percent had a stated GHG reduction goal, only 5 percent had taken the step of performing a GHG inventory, developing a reduction plan, or giving climate change a designated budget. In Tozer's (2013) assessment of Community Energy Plans in Canada, she finds that city governments have struggled to implement the programs and projects needed to reach their GHG emission targets. Only one city, Vancouver, has implemented programs in each of the sectors targeted by their plan.

Even in California, where the state government requires local governments complete a Climate Action Plan, implementation lags. A survey of thirty-four large cities in California conducted by students at the University of California, Santa Barbara shows an average implementation rate of 35 percent; and only one-third of the cities were able (or perhaps willing) to report the emission reductions they had achieved as a result of their efforts (Qin et al. 2014).

A final example of how cities are progressing can be found in the 2010 ICMA Local Government Sustainability Survey, which asks more than 2,000 U.S. city governments about the environmental programs and policies they have in place. The survey shows that while more than half of these cities identify climate change as a priority, less than 10 percent have a GHG emissions inventory and only 44 percent have even taken the step of installing energy efficient lightbulbs in city-owned buildings.

Ambition has not necessarily always led to progress. While there is clearly evidence of climate change mitigation moving forward in cities, assuming the policy ambitions of city governments will translate to comprehensive and far-reaching changes to urban systems seems premature. The "stubborn gap between the rhetoric and reality of local climate policy" identified by Betsill and Bulkeley a decade ago (2007, 448) continues today. We know that cities are ambitious, and we know some of their reasons for taking on this challenge. But how can we better understand the gap between their ambitions and progress and, more importantly, move toward filling it?

The Focus on Factors and Barriers

The gap between cities' ambitions and progress has not gone unnoticed by scholars. A small but growing body of research grapples with the question of what leads to more or less success in climate change mitigation in cities. Rather than emphasizing the features of the implementation process or the nature of the governing strategies that matter, a significant portion of the work has focused on identifying the *characteristics* of cities that are associated with more progress and the barriers cities face as they move toward implementation. While this research has produced new and very important insights, it also leaves us with various lists of "factors that matter" for the success of city governments (Ryan 2015) and no clear point of departure for interpretation.

While the growing list of factors that matter is a useful starting point, the findings have been inconsistent and inconclusive. For example, a number of studies have found that city governments with greater internal capacity (whether legal, organizational, financial, or technical) have more success in meeting their mitigation goals (Zahran et al. 2008; Bulkeley 2010; Krause 2011a; Sharp, Daley, and Lynch 2011). Conversely, a lack of capacity is often cited as a reason for failed or stalled implementation (Betsill 2001; Burch 2010; Aylett 2013; Romero-Lankao et al. 2013; Homsy and Warner 2015). It is also clearly the case that many traditionally high-capacity cities struggle to make good on their climate change commitments (Burch 2010). The relationship may be conditional on other factors. For example, research on implementation progress in C40 cities, the set of large global cities that have committed to reducing GHG emissions, shows that a city's GDP is only related to implementation at the extremes (ARUP 2014).

Scholars also emphasize factors such as political leadership, jurisdictional autonomy, or the socio-demographic makeup of the city (Betsill 2001; Romero-Lankao 2007; Bulkeley 2010; D. R. Pitt 2010; Sharp, Daley, and Lynch 2011). The empirical record and theoretical development around these factors are far from conclusive as to exactly how and when these factors matter for climate change mitigation in the city and how important they are in relation to one another. For example, Bulkeley (2013) considers leadership to be "a necessary, but not sufficient, factor in developing sustained climate-change-mitigation action" (p. 136–137). While clearly there are conditions that make it easier for cities to act, we still need to understand the processes by which cities take advantage of their opportunities and avoid potential pitfalls.

A related body of work has sought to identify the barriers cities face in making progress on climate change mitigation and has produced a growing list that mirrors the factors for success: financial constraints, jurisdictional constraints, a lack of technical capacity, and a lack of political leadership or public support (e.g., Betsill 2001; Bulkeley and Betsill 2003; Robinson and Gore 2005; Clean Air Partnership 2007; Tozer 2013). Scholars have also emphasized the barriers cities face in their broader policy context for facilitating or constraining climate policy choices. For example, Bulkeley and Kern (2006) found that in the early 2000s cities in Germany and the UK faced shared structural and political challenges in their mitigation efforts that made it difficult to tackle community-wide emissions. Decision makers are often asked what barriers they perceive to their efforts to reduce GHG emissions, and they typically provide a long and relatively predictable list that includes financial and political constraints (e.g., Tozer 2013; Aylett 2014).

The empirical relationship between perceived barriers and the progress of city governments in addressing climate change is not typically examined, and when it is the results are ambiguous at best. For example, the survey of thirty-four large cities in California referenced earlier also included information about the barriers city governments perceive in their efforts to implement their climate action plan (ranging from funding to training to local opposition) and the percentage of their plan they estimate to have implemented (Qin et al. 2014). As figure 1.2 shows, there is no discernible relationship in cities between perceived barriers and progress on implementing the climate change plan. In fact, on average these cities feel relatively neutral about the significance of such barriers. While this is not meant to be a definitive test of how administrative and political barriers relate to implementation progress, it does provide further evidence of a need for more nuanced explanations of whether and how such barriers impede progress on urban climate change mitigation.

A focus on factors and barriers reemphasizes that cities have taken on a difficult task in committing to address climate change. We know city governments will face political, institutional, and financial challenges, as these are inherent to the task of governing GHG emissions. Delineating the kinds of factors and barriers cities specifically face helps us to understand what cities are starting with but not necessarily where they are going. This approach also makes it difficult to provide decision makers with support and advice as there is little that can be done about a contentious state political environment or a population with lower levels of education. To understand and support climate change mitigation progress in

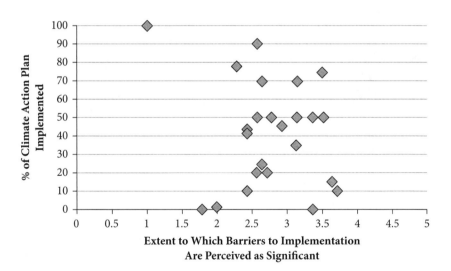

FIGURE 1.2 The relationship between perceived significance of barriers to reducing GHG emissions and progress in implementing a climate action plan for twenty-five California cities. X-axis measures average response from city governments to statements that eighteen different barriers were significant. 1 = Strongly Disagree, 2 = Disagree, 3 = Neutral, 4 = Agree, 5 = Strongly Agree. Data drawn from Qin et al. 2014

cities, we need to move beyond these initial insights and toward a greater understanding of how city governments act in the face of complexity.

The Need to Ask "How?"

Significantly reducing urban GHG emissions requires that city governments successfully navigate a new and highly complex policy landscape, one with elements that lie largely outside their control. Good progress has been made in identifying the types of factors and barriers that can parameterize policy choice and shape decision making. Such work demonstrates empirically that reducing urban GHG emissions is fraught with multiple sources of uncertainty and a host of challenges. It provides a solid foundation from which to interpret the ambitions and progress of city governments on climate change mitigation.

These approaches fall short as explanations of how city governments can, do, or should navigate this complex landscape. We are left with a list of things city governments may need—authority, resources, data, leadership—but without a sense for how (or how much) these things matter and how city governments can get them. These approaches also fail to specify the processes at play in addition to the factors in place. They emphasize the role of jurisdictional constraints, financial strain, and multilevel dynamics without clarifying the role city governments do play in such a complex landscape and how they can best leverage this role to achieve their climate change goals. How can cities navigate the gap between ambition and progress? What are the governance practices that allow city governments to overcome barriers to action? As city governments take up the mantle of climate change mitigation leadership, should we be optimistic?

Climate change mitigation is a hard problem for cities, but their success is crucial. Despite the high political and ecological stakes of climate change mitigation in cities, the governance dynamics involved in moving from a GHG reduction target to effectively developing and implementing the necessary policies and programs have been largely unexamined and under theorized. There is unlikely to be an ironclad list of factors that predicts urban climate change mitigation success. We must move beyond the challenges present in urban climate change mitigation and understand the processes by which city governments work through and navigate complex landscapes. A new approach is needed if we are to understand the mechanisms and strategies that underlie the work of city governments to take their cities in a new direction. In short, we need to understand how to repower cities.

Gaining traction on the processes of urban climate change mitigation requires a framework that accounts for both the complexity city governments face and the strategies employed to steer the city in a new direction. It also requires detailed empirical work on the inner workings of cities and urban governance processes. We must begin to trace the journey cities take from setting an ambitious target to realizing the social and technological changes needed to reduce GHG emissions; to move from questions focused on why cities are acting and what might prevent their success, to questions focused on the processes and mechanisms by which climate change mitigation takes place in cities.

In the next chapter I develop a framework that accounts for the complexities involved in reducing GHG emissions and identifies a set of strategies city governments can and do use to navigate this complexity. In doing so I also take up more fundamental questions about the role of city governments in shaping their cities, the strategies available for mobilizing needed resources and actors, and the means by which we should be evaluating their success. I argue that the factors and barriers that have been highlighted in previous research matter more for the path cities ultimately take than for whether they make progress on reaching their destination. There are governing strategies city governments can use to reduce uncertainty, mobilize actors and resources, and fuel their efforts to take their city in a new direction. These strategies prove crucial to the work of New York City, Los Angeles, and Toronto to reduce urban GHG emissions and provide an entry point for identifying approaches that will take cities from first steps toward climate change mitigation to true urban transformation.

2
EVALUATING URBAN GOVERNANCE
A Three-Part Framework

Understanding the implications and potential of urban climate change leadership requires revisiting and specifying the role of city governments in shaping their city's trajectory, the choices they make about policy tools and governance strategies, and the means by which we should be evaluating their success. Governing climate change mitigation is a matter of not only setting a climate change mitigation target but also mobilizing and directing the vast array of people, investments, and activities that produce urban GHG emissions toward the collective goal of climate change mitigation. As Vanessa Castán Broto points out, "these are not just well-intentioned means to respond to an urban problem, but complex political entanglements which reveal the underlying paradigms and the contexts of intervention in which governing actors operate" (2017, 8). Reducing GHG emissions represents an entirely new pursuit for cities. Perhaps as a result of its uncertainty and complexity, the process of moving from an emissions reduction target or climate change plan to real change in the city is something of a theoretical and empirical black box.

As city governments take on more complex policy challenges, including climate change mitigation, new approaches are required for understanding their pursuit and its impacts. Jill Simone Gross captures this gap nicely, finding that "where we are lacking is in the adequate development of research and theory that focuses specifically on these processes and the points at which hybrid collaborations lead to governance transformations (structurally or institutionally or in outputs)" (Gross 2017, 561). In this chapter I parse out the key elements of urban climate change governance, and forward a framework that specifies the relationship between context and policy, the use of key governance strategies by city governments, and the means of evaluating the progress and contributions cities are making to climate change mitigation.

The Role of City Governments in Urban Governance

In the past, the role of city governments, their policy priorities, and their ability to shape the city have been understood as determined by the strength and exercise of their formal powers (Kjaer 2009, 139). City governments were often presented as limited and relatively neutral, or at least politically insulated, policy implementers who exerted the direct influence they had on other actors in the city through their use of traditional municipal policy tools: zoning and land use policy, building codes, tax rates, user fees, regulation of local industries (taxis, restaurants, etc.), permitting, and directing infrastructure investment (e.g., Peterson 1981).

There are reasons to take issue with this characterization of urban policymaking and politics. Cities are indeed constrained in their formal authority and resource availability, but the result is not necessarily limited or politically insulated city government. Rather, the constraints on city governments mean they govern through complex networks of authority and agency (Jones 1983; Hambleton and Gross 2007). Critical economic, technical, and political resources necessary to shape—or reshape—urban communities and economies are held by the private sector, civil society, and other levels of government (Stone 1989; Pierre 2014; Stone 2015). Urban governance landscapes also extend beyond the city and change over time, as the priorities and actions of other levels of government, national economic trends, and even international and transnational organizations all have the potential to add or remove investment and capacity, reconfigure local political coalitions, and influence the preferences of urban decision makers. The power of city government comes not only from formal authority or electoral mandate ("power over") but also from an ability to produce results and to mobilize others ("power to") (Stone 1993, 24). The challenge for city governments is to develop strategies for accessing, mobilizing, and coordinating these resources and actors.

Urban regime theory and other political–economic approaches to urban politics emphasize the need to activate the capacity of the local state to govern by mobilizing and coordinating a diverse array of participants and fragmented political and financial resources (e.g., Stone 1993; Jessop, Peck, and Tickell 1999; McGuirk 2003). They recognize explicitly that political power and institutional capability is less and less derived from formal constitutional powers accorded the state but more from a capacity to wield and coordinate resources from public and private actors and interests. Urban governance theory similarly departs "from a notion of insufficient institutional capabilities to allow the city to perform key roles in governing" (Pierre 2014, 873) and emphasizes the emerging role of city governments as a "network governor" as they navigate increasingly complex webs of authority. Likewise, many policy implementation scholars have shifted their focus to the broader question of "how systems of governance deliver policy-relevant impacts" (O'Toole 2000, 276).

Existing theoretical approaches typically fail to account for the conditions under which cities might *succeed* in pursuing strategies that do not explicitly target growth (Mossberger 2009). Urban regime theoretical approaches emphasize the limitations of city governments rather than how or whether these limitations might be overcome in pursuit of a collective aim. It does not specify whether or how city governments might build a coalition able to respond to climate change or what other governance changes might be necessary to pursue climate change mitigation, especially given the uncertainties it presents. Urban governance theory lacks specificity as to the role of city governments, providing anything from "state-centric models of institutional control" to "extensive network governance with the local state mainly playing the role of a metagovernor" (Torfing et al. 2012; Pierre 2014, 866). Urban governance presents a broader landscape of actors and structures, but again no clear expectations for how city governments might effectively navigate this landscape in pursuit of a transformative agenda.

Reducing urban GHG emissions has as much to do with reorienting the development trajectory of a city as it does with implementing a particular policy or program (Burch et al. 2014). Cities need to be placed on an entirely new path. This requires assembling and enabling collective action, along with proactive leadership and creative exercise of authority on the part of city governments. While the landscape of urban politics is fragmented, and the resources needed to govern the city are distributed, city government can play an important leadership role in channeling and coordinating these resources. Even in their context of diffuse and fragmented authority, city governments have the ability to set goals or a direction for the city, provide accountability and legitimacy to decision-making processes, and coordinate the actions of public institutions, market actors, and voluntary associations to ensure that decisions are implemented (Stoker 1998a; Andrew, Graham, and Phillips 2002; Savitch and Kantor 2002; Pierre and Peters 2012; Rich and Stoker 2015). City governments can have a more or less central position in the resulting governing arrangements (Clarke 2016), but in pursuing a transformative agenda (such as climate change mitigation) their purpose should be to mobilize and coordinate participation and resources from within and outside of government toward this aim.

Placing the focus on the choices and strategies of city governments acknowledges the growing tendency of city governments to engage in global problem solving and set an ambitious agenda for the city absent federal direction or incentives. Indeed, North American cities can no longer rely on strategic direction or significant financial support from federal governments. As Jon Pierre (2011, 148) writes, "the era of federal urban policy is, like, way over . . . Today, the prosperity of cities hinges on their ability to formulate strategies and mobilize resources for their development . . . the future of a city or region is now much more up to itself than was previously the case." Further, no other actor in urban governance has the legitimacy, authority, or incentive to pursue an ambitious collective goal like climate change mitigation. City governments set a direction for the city, provide accountability and legitimacy to decision-making processes, and coordinate the actions of public institutions, market actors, and voluntary associations to ensure that decisions are implemented (Stoker 1998a; Pierre and Peters 2012; Rich and Stoker 2015). In the face of powerful, complex, and global forces, city governments navigate their environment very strategically and are able to shape their city's trajectory (Clarke and Gaile 1998; Savitch and Kantor 2002; Sellers 2002a). City governments have the ability to make choices about how to best coordinate and motivate key actors and resources: when to use direct intervention, how to engage the public, where their own resources will be directed, and who to partner with at what stage in the process. Such choices are critical for mobilizing partners and resources and serve to "activate specific powers and state capacities inscribed in particular institutions and agencies" (Jessop 1990, 366–67).

The formal authorities and jurisdiction of city governments are opportunities to directly shape behavior and can also be used for gaining access to resources controlled by private businesses or other partners; but formal authority alone is not what constitutes or indicates the governing capacity of city governments or their ability to bring about change. City governments must work actively to generate the capacity to govern, regardless of the formal authorities they may hold. They must work in partnership with those who control the disparate resources required to solve a problem (Pierre 2014) or whose participation is required for change to take place. City governments can work through partnership networks to achieve their aims: they can give authority to networks or create new networks (Peters 1997).

Deploying these strategies is akin to what Osborne and Gaebler refer to as *steering*: "governments that focus on steering actively shape their communities, states, and nations. They make more policy decisions. They put more social and economic institutions into motion. Some even do more regulating. Rather than hiring more public employees, they make sure other institutions are delivering services and meeting the community's needs" (Osborne and Gaebler 1991, 32).

It is not immediately obvious that city governments are well practiced in some of the more novel "steering at a distance" strategies of urban governance (Healey et al. 1995; Pierre 2011), but such strategies are increasingly required for progress on government-led agendas (McGuirk 2003).

While perhaps an uncomfortable claim in the current era of governance and networked politics, analytically positioning city governments in the driver's seat helps us to better understand the space between a government-led target or ambition—like reducing urban GHG emissions by 80 percent—and its outcomes. The key lies in being clear about what occupying the driver's seat entails and how our understanding of this position has changed over time. The approach city governments take will be different in different kinds of cities—the modes of governing they work through will vary, for example—but all city governments have a central role to play in governing toward a particular, predetermined purpose or outcome such as climate change mitigation. This is distinct from urban governance on autopilot, or decision makers who are merely responding to external conditions or structural imperatives.

Participant mobilization and resource coordination are the mechanisms by which cities make progress toward a collective aim and are ultimately what constitutes the role of city governments in urban governance. In specifying this task, I aim to shift the theoretical and empirical emphasis from the structure or composition of governing arrangements to the actions of city governments, and specifically the ways in which they leverage authority to pursue a progressive and ambitious collective aim. Foregrounding city governments reveals how they can and do play an active role in mobilizing the resources and actors necessary to take their city in a new direction and the ways in which this might play out the same or differently in cities with very different political, technological, social, and institutional characteristics. It positions the choices and strategies of city governments as the analytical centerpiece, and it subsequently provides insight into how city governments not just govern their cities but place their cities on a new path altogether.

City Governments and the Challenge of Climate Change Mitigation

Committing to a GHG reduction target, or even developing a plan for addressing climate change, is a relatively low-cost policy action for city governments; taking steps to implement the policies and programs necessary to meet those reduction targets is where the real challenge lies. Climate change mitigation presents decision makers with the twin challenges of complexity and uncertainty.

TABLE 2.1 Climate change mitigation policies and programs for the energy, transportation, and waste management sectors. Adapted from IPCC 2014

MITIGATION OPTION	ENERGY SECTOR EXAMPLES	TRANSPORTATION SECTOR EXAMPLES	WASTE MANAGEMENT SECTOR EXAMPLES
Consume Less	Turn lights off in buildings; avoid the need for heating and cooling systems	Reduce transportation demand; reduce distance of trips	Reduce waste production; improve collection routes in cities
Consume Using Less Energy	Use energy efficient appliances, light bulbs, and HVAC systems; improve efficiency in energy generation	Increase public and active transit options; increase vehicle fuel efficiency	Waste-to-energy schemes including incineration and gasification
Use Low-Carbon Energy Sources	Introduce renewable energy sources (i.e., wind and solar) or natural gas	Use electric vehicles or biofuels	Capture methane emissions at landfills

The complexity of urban climate change mitigation stems from the fact that, to reduce GHG emissions, city governments must engage with fundamental infrastructures of energy and mobility and the social practices that shape their use. This implicates nearly every sector and requires the coordination of a range of urban actors, from residents to city agencies to national governments (Chester et al. 2014; Marcotullio et al. 2014). The City of Toronto, for example, warns that reducing the city's GHG emissions by 80 percent "will require big changes in the ways that we live, travel, work and consume goods and services, which will in turn affect everything from the structure of our local economy, to the types of jobs we hold, to our education and training programs, and more" (City of Toronto 2016a). There are three broad categories of climate change mitigation options: consuming less, consuming in ways that use less energy, and using low-carbon energy sources (IPCC 2014b, 397). Table 2.1 provides some examples of how these categories are operationalized for energy, transportation, and waste management systems in cities.

A recent survey of C40 cities found that North American cities are most likely to govern climate change through a mix of direct powers and collaborations and partnerships (ARUP 2015). Many mitigation projects and programs will require city agencies to be willing to work together, across traditional jurisdictional boundaries, in pursuit of objectives that might deviate from traditional mandates. Energy supplies may be fully or partially governed by private utilities; state, provincial, or federal governments; or other municipalities. Transportation networks are perhaps even more complex as they are typically funded, built, and operated by a mix of public and private actors. Subsequent mobility patterns (and GHG emissions) are determined by an additional set of choices about work, housing, and recreation opportunities. The behaviors of building occupants (e.g., opening windows, leaving lights on, setting thermostats) play a significant role in shaping energy use, and architectural solutions alone are not sufficient to meet mitigation targets (Haldi and Robinson 2011; Janda 2011). Private sector participation is needed for retrofits that reduce energy use in commercial and industrial buildings, and for investment in alternative development patterns (such as transit-oriented development) that facilitate lower GHG emissions. Civil society assists in implementation and providing technical expertise. Coordination with other levels of government—state/provincial and federal regulating and granting bodies—and the private sector can be necessary for harmonizing standards and acquiring needed funding.

Beyond the complexity presented by climate change mitigation, city governments must grapple with significant uncertainty. Climate change mitigation is emblematic of what have been called "wicked" policy problems, meaning the definition and scope of the problem is often contested, there is not always a clear set of potential solutions, and it has uncertainty at its core (Rittel and Webber 1973). Koppenjan and Klijn (2004) identify three types of uncertainty characteristic of wicked problems—substantive, strategic, and institutional—that are useful for understanding the challenge of urban climate change mitigation.

Urban climate change mitigation involves substantive uncertainty, or uncertainty about the nature of the problem itself—the extent and drivers of GHG emissions in a city—as well as political contestation over how such information should be interpreted. Most cities are only beginning to collect baseline data describing their GHG emissions. Such data typically include information about energy generation, energy consumption, and transportation patterns. Cities with private energy utilities may not be able to easily access information about energy consumption due to privacy considerations and utilities' concerns about the regulatory implications of greater transparency. Transportation data might only be collected on a sporadic basis, or as part of five- or ten-year planning cycles, and at regional rather than municipal scales. Without robust datasets it can be difficult to determine where a city's GHG emissions come from, what the trends have been over time, and how they might vary geographically within the city.

Even when data are available, city governments may lack the technical capacity to conduct a comprehensive GHG emissions inventory and can use very different inventorying methods and tools. Accurate accounting of urban GHG emissions is challenging, as it is not always clear what the geographic boundaries of a city's GHG emissions should be. Electricity grids and transportation networks are often regional in scale, and city governments have varying levels of authority over important sources of GHG emissions such as aviation and shipping. In an effort to clarify the boundaries of urban GHG emissions, a distinction is typically made between what are called Scope 1, Scope 2, and Scope 3 emissions. In the urban context, Scope 1 emissions are those coming from sources located within the city boundary; Scope 2 emissions as those coming from the use of grid-supplied electricity, heat, steam, and/or cooling within the city boundary; and Scope 3 emissions as those emissions occurring outside the city boundary as a result of activities taking place within the city boundary (Fong et al. 2015). Choices about what to include or exclude from an emissions inventory can significantly alter the final tally (Ramaswami et al. 2008). As described earlier, efforts have been made to standardize the inventorying process, but implementing these requires capable and available staff, which is not available in all city governments.

Substantive uncertainty also refers to the political contestation that often accompanies the production and interpretation of new climate change mitigation information. It can be difficult to secure shared definitions of the problem of climate change (i.e., as an environmental, economic, or security problem), and it is unclear who should bear the responsibility for its consequences. For example, knowing the per capita GHG emissions of a city does very little to resolve claims about how to best reduce these emissions or even whether there is a compelling reason to do so. Observing that a city's GHG emissions have shifted over time from industrial energy use to transportation systems similarly does not provide guidance on how policymakers will or should interpret this shift. Disagreement over the definition of the climate change problem for cities runs deeper than data and detailed inventories, and it contributes to the wicked nature of urban climate change mitigation.

Urban climate change mitigation is also characterized by "complex and highly undetermined types of interactions," emblematic of strategic uncertainty in wicked problems (Koppenjan and Klijn 2004, 7). Urban infrastructure and behavior patterns shape and reflect the foundation of urban economies and communities, as the development and maintenance of most cities is underpinned by a reliance on fossil fuels (McNeill 2000; Smil 2008; Marcotullio et al. 2014). Indeed, these systems are fundamental to how societies operate and function (Barr et al. 2014), and changing them requires an ability to overcome "lock-in" and "path dependencies" of previous policy decisions and investments (Unruh 2000). Urban GHG emissions are the product of a complex mix of embedded social, political, and economic practices and institutions, and mitigation requires large-scale buy-in and behavior change. There is significant uncertainty about what effective, efficient, and equitable solutions look like, how to pursue them, and who is most likely to engage.

Climate change mitigation introduces new politics into old systems and can deepen ideological rifts in many contexts. Participants are likely to have different objectives and interests, and actors will make a variety of strategic choices about how to articulate the problem of urban climate change mitigation. As Koppenjan and Klijn point out, with wicked problems "it is difficult to predict what strategies actors will choose and how the 'interaction' of these strategies will influence the problem situation and problem solving process" (2004, p. 7). Choices in one sector can undercut or displace the mitigation efforts of another.

Similarly, ambitious GHG emission reduction projects that require new or reconfigured infrastructure, retrofitted buildings, or accumulated behavior changes can take years or decades to complete. In some cases, efforts to reduce emissions may primarily benefit future generations, but they require that decision makers prioritize them in the short term. There is deep uncertainty in our understanding of how various systems and choices are connected to one another in ways that matter for urban GHG emissions, the costs and benefits associated with different approaches to mitigation, and the strategic choices of different actors in the system.

Finally, reducing urban GHG emissions crosses sectoral and jurisdictional boundaries, emblematic of institutional uncertainty in wicked problems. Unlike many other policy areas, urban climate change mitigation requires the coordination and buy-in of multiple sectors, from energy to housing to transportation. Other levels of government, actors from the private sector, regional bodies, and civil society organizations are also implicated in urban climate change mitigation. These groups will be differentially "guided by the tasks, opinions, rules and language of their own organization, their own administrative level and their own network. . . . For all actors at the beginning of the process, this results in a high degree of uncertainty about how the process will be handled and how the interaction with other actors will develop" (Koppenjan and Klijn 2004, 7).

Authority can also be highly fragmented in ways that constrain a given city's ability to implement mitigation programs and reduce GHG emissions. Metropolitan or regional fragmentation can inhibit local action on issues such as transportation, energy sourcing, and land use due to conflicting aims and resources held by other cities (Feiock 2004; Judd and Swanstrom 2012) by generating disincentives and transaction costs that may stifle a city government's agenda.

In many cases climate change mitigation represents an entirely new set of tasks for decision makers and managers in city governments—for example, installing solar panels or encouraging private building owners to conserve energy—and in other cases introduces a new set of decision criteria, such as the GHG implications of new transportation projects. It is not always clear how to reconcile these new tasks and criteria with existing institutional arrangements and organizational mandates.

City governments grapple explicitly with this institutional uncertainty as they develop their climate change programs and policies. As described earlier, most early mitigation action in cities focused on reducing GHG emissions from city government operations, a problem space with significantly less institutional uncertainty (Betsill and Bulkeley 2007). As city governments have gained confidence and knowledge, their focus has expanded to the GHG emissions of the broader community. Nonetheless, operating in this institutionally complex landscape presents a relatively novel task for city governments. Further, navigating institutional uncertainty may be particularly challenging at the local level, where city governments are often working with limited jurisdiction and authority within multilevel governance systems.

In the face of this complexity and uncertainty, reducing urban GHG emissions requires intervention and leadership from city governments and strategies for mobilizing and coordinating the diverse landscape of actors and resources. Climate change mitigation is not a "spontaneous social product: it requires goal-oriented intervention by governments and other actors" (Meadowcroft 2007, 302). While important initiatives from outside of city government can help to spur action and raise awareness (Bulkeley, Broto, and Edwards 2014), ultimately it must be the city government that changes policy, legitimizes and advances the cause, and provides the needed financial and regulatory incentives for transformative change. Residents, civil society, and the private sector look to the city government for leadership and guidance. As a leader of a nongovernmental organization in Los Angeles put it, "if the goal is activating action, guidelines are helpful . . . an absence of guidance from the City (on climate change mitigation) would be worse than poor guidance."[1]

Indeed, as discussed previously, scholarship has shown consistently that city governments are rarely so constrained that they are left without any type of agency or choice regarding how to pursue their goals. City governments have multiple forms of authority and power when interacting with the private sector, civil society, and other levels of government. In short, the choices and strategies of city governments matter. City governments operate in a complex and multi-scalar policy environment, but they are constantly making choices about how to interact with, or even shape, that environment. They are highly strategic in their navigation of broader policy contexts, though some strategies will be more effective than others (Clarke and Gaile 1998; Sellers 2002a). While there are barriers and structural factors that will shape how cities engage, they do not necessarily determine the outcome. For example, in Rich and Stoker's (2015) examination of the role of local governance in determining the successful implementation of federal Empowerment Zone programs in the United States, they find that "poor local governance can squander a favorable context and good local governance can overcome a difficult context. This suggests that while contextual factors are important, what cities do in response to local context matters" (Rich and Stoker, 2015, 7).

The governing task, then, for city governments pursuing deep GHG emission reductions is to mobilize and coordinate the participants and resources needed to meet new collective aims, to restructure and redirect the networks of influence and engagement in the city in ways that foster transformative change. In addition to the participation of multiple actors, financial, political, and technical resources must be assembled and coordinated in order to make changes that reduce GHG emissions, such as expanding a public transportation system or rewriting the city's building code to require energy efficient design elements. Financial resources are required to pay for new infrastructure, hire new staff, produce new reports, and implement new programs that the city will need to reduce GHG emissions.

City governments are unlikely to hold, or be able to generate themselves, the financing necessary for achieving significant GHG emission reductions. Changing energy sources (e.g., installing thousands of solar power panels) or transportation infrastructure (e.g., building a new subway line) requires significant capital investment. City governments often have a prescribed set of revenue-generating tools and strict caps on the amount of debt they are allowed to carry. Financing from other levels of government or the private sector (in addition to municipal financing) will likely be necessary. Therefore, city governments will need to not only mobilize their own internal financial resources (including committed staff support) but also actively find ways to solicit, attract, borrow, or otherwise secure outside investment if they are to meet their GHG emission reduction goals (Peterson 2018).

Urban climate change mitigation requires mobilization and exchange of political resources. City council may need to adopt bylaws that are not popular with the development sector or with certain populations in the city. Mayors will need the support of their city councils (and vice versa) to pass legislation that supports an emissions reduction target. Decision makers must demonstrate leadership and political courage and negotiate support from the business community. For example, the city might choose to set higher energy efficiency standards in new buildings or to adopt a tiered energy pricing scheme that charges large energy users more. While such actions are a step toward reducing GHG emissions, they are likely to be unpopular. Political support from the public and private sectors to pass such bylaws must be generated and political interests aligned in ways that allow such measures to be effective.

Cities will also need to mobilize and coordinate technical resources to support their efforts to reduce GHG emissions. Cities must track their GHG emissions over time if they are to demonstrate their effectiveness in meeting their targets. This requires emissions data from a range of sources that can include nearly every city department as well as state and national transportation departments, privately owned energy utilities, and private waste haulers. Finding and using these data can present a significant challenge for city governments. City governments may also be aided by new technologies and technical capacities, such as energy efficient building technologies and renewable energy options compatible with the city's existing portfolio.

The key question is how this task of governing cities is accomplished—how the agency of city governments is put to work—and specifically how it can be accomplished by city governments pursuing major urban transformations such as climate change mitigation. As a step toward answering this question, I develop a framework for structuring analysis of *how* and *where* city governments govern their cities as they seek to reduce GHG emissions, allowing for deeper understanding and evaluation of the tools available to and deployed by city governments.

The framework has three components. First, city governments make choices about the policies and governing modes they will use to reduce GHG emissions. These choices represent each city's unique route to climate change mitigation and are shaped by the broader social, political, institutional, and physical context. This first-level *how* of urban climate change governance comprises the unique set of programs and policies used, and the mode(s) of governing this set represents—from more to less direct intervention. Second, regardless of the specific route a city chooses there are shared governing strategies city governments can and do use to mobilize participants and resources: institution building, coalition building, and capacity building. These strategies allow city governments to reduce key sources of uncertainty, mobilize the participants, and coordinate the resources needed for change. They underpin and activate the ability of city governments to govern climate change. Third, evaluating urban climate change governance requires evaluating its impacts. These are both reductions in city-scale GHG emissions and broader changes in the city, and beyond, catalyzed by efforts to reduce urban GHG emissions. Given the diversity of emission sources and solutions between cities, there are likely to be multiple routes by which cities achieve similar impacts. City governments can make progress on climate change mitigation, even when they lack formal authority or face resource constraints, by making strategic choices about how to mobilize needed resources and participants.

Part 1: Policy Agendas Tailored to Context

Cities embedded in different systems, imbued with different sets of formal and informal authorities, are likely to face unique opportunities for channeling collective action (Stone 1993, 17; Sharp 1997, 285; Pierre 2011, 106). City governments confront a unique physical, political, and economic landscape armed with an equally unique set of opportunities and priorities. This context sets the legal, technical, and political parameters in which city governments must work and make their decisions. The relevant features of a city's climate change mitigation context reflect the factors and barriers discussed in chapter 1: their sources of GHG emissions, formal authorities, capacities, leadership strength, and larger policy context.

Rather than determining whether city governments are likely to intervene to reduce GHG emissions, or even whether they will be successful, this context shapes the policy agenda cities ultimately develop to mitigate climate change. The choices city governments make about their agenda are reflected in the sectors they target and the mode of governing they use. The opportunities and obstacles a particular city faces when pursuing GHG emission reductions shape these choices.

Choice of Sectors

City governments can choose to focus their mitigation efforts on one or more contributing sectors and to shift this focus over time. Energy supplies, energy demand, transportation infrastructure and demand, and waste production and management are all potential intervention points for reducing urban GHG emissions. The choice of sectors can be shaped by a number of factors embodied in the city's context.

A distinctive network of technologies, infrastructures, actors, and practices supports the mix of sectors and energy sources underpinning a city's GHG emissions. This network includes the "material basis of urban climate governance" (Bulkeley 2010, 147), such as hard infrastructure underlying energy and transportation systems, as well as citizen and private sector support, state and federal support programs, and the extent to which climate change mitigation programs align with existing goals for the city. These features help determine the financial, political, and technical resources required for emission reductions.

In choosing to target different sectors, city governments are choosing to take on different governance tasks. Changing transportation and mobility patterns in the city is a much different governance task than reducing energy use in buildings. The former may require large capital investments and intergovernmental coordination to improve public transportation, while the latter may require voluntary behavior change and a large number of small retrofitting projects to increase efficiency. Similarly, there are different actors, regulations, and technologies at play in shaping energy demand and energy supplies. Significantly reducing energy demand in privately owned buildings in a city requires that thousands (or millions) of individuals make different choices about the technologies and practices they use to heat, cool, and electrify their homes and offices. Homeowners may need to upgrade windows, use energy efficient lightbulbs and appliances, and turn off lights and HVAC systems when they aren't needed. By contrast, removing coal from the city's energy supply might require the renegotiation of a contract between the city and its energy supplier. Both options are likely to reduce GHG emissions, but targeting different dimensions of the energy sector presents unique demands on city government, engages different institutional mechanisms for implementation, and implicates a very different set of stakeholders.

The scope of authority held by city governments over its GHG emissions varies from source to source, between countries, states, and provinces, and in some cases over time. Cities may choose to focus on sectors over which they have greater authority, particularly in the early phases of their work. A survey of C40 member cities found that city governments have less direct control over their energy supplies relative to other sectors; only 23 percent have control over centralized power generation (ARUP 2014). A city government that controls its energy system, like Los Angeles, may be better positioned to reduce GHG emissions through the introduction of renewable energy supplies. Alternatively, a city government with significant authority over building codes, like New York City, may be better positioned to reduce GHG emissions through energy conservation design criteria.

Similarly, cities may focus on areas where they have greater levels of capacity, either through existing funding streams, previous experience, strong staff or public support, or technical expertise. A city with a strong green building program in place might find it advantageous to focus on reducing building energy use, while a city with a strong biking or public transportation culture might choose to focus on expanding those programs. Different departments within city governments might play a leadership role and therefore shape priorities as well. In Toronto, the city's Finance Department was eager to contribute and developed a number of innovative financing mechanisms for energy efficiency retrofits.

Given the prominent role mayors have played in getting climate change on urban policy agendas, the powers and priorities of a city's mayor can determine where opportunities for intervention are most likely (Clarke and Gaile 1998; Greasley and Stoker 2009). A vocal and committed mayor can help prioritize climate change mitigation, particularly in ways that align with his or her broader policy agenda. For example, former New York City mayor Michael Bloomberg emphasized the energy conservation opportunities in the private sector, while Bill de Blasio is focusing on opportunities to use climate change mitigation measures to increase housing affordability in the city. City governments may choose to prioritize those sectors where the mayor has greater influence or interest. There is tremendous variation in the way that authority is allocated within city governments, and mayors are given different powers over budgets, appointments, and the legislative process (Mullin, Peele, and Cain 2004).

The choice of sectoral focus can also be shaped by the broader policy context in which city governments are operating. State and federal policies and funding programs can significantly shape the opportunities local governments have to innovate and implement new initiatives (Sellers 2002b; Frug and Barron 2008). The choices made by other levels of government can support certain goals and even provide additional resources and opportunities for action in particular sectors. Alternatively, other levels of government can generate disincentives and increase transaction costs in ways that stifle a city government's agenda. For example, a state or provincial government might shift financial incentives from the development of renewable energy sources to energy conservation or climate change adaptation. In Toronto and Los Angeles, provincial and state prioritization of solar energy has helped to propel local investments in these technologies as well.

The political economy of energy systems can shape the priorities of cities. Cities rely on unique mixes of fossil fuels, including natural gas, heating oils, and coal, largely determined by regional or even national energy markets and networks and with widely varying GHG emission profiles (figure 2.1). A survey of C40 member cities shows that fuel sources for heating buildings vary significantly by global region: oil represents around 60 percent of heating fuels in North American cities but only around 25 percent in European cities (ARUP 2014, 113). This mix helps to determine the CO_2 intensity of a city's energy supply, but it also structures the intervention opportunities for city governments. Los Angeles was almost entirely dependent on coal in 2007, providing huge opportunities for GHG emission reductions in their energy supplies. Toronto and New York City were

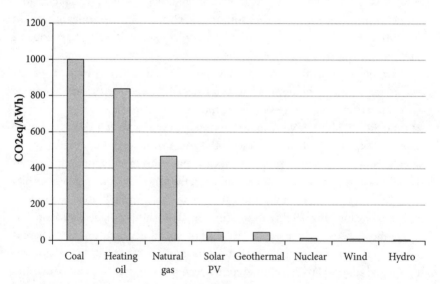

FIGURE 2.1 GHG emissions per kilowatt hour of renewable and nonrenewable fuel sources (median life cycle). Adapted from the Intergovernmental Panel on Climate Change (IPCC 2011)

more reliant on natural gas, meaning the emission benefits from renewable alternatives were relatively smaller.

National economic conditions can also help to determine the choices of city governments. Large pots of money for transportation projects or energy conservation or community development would help to incentivize a different set of GHG emission targets and programs; an economic recession or sudden spike in unemployment may lead cities to reevaluate their approach altogether. The 2008 recession led Los Angeles to do away with its Environmental Affairs Department all together. Market conditions such as housing prices and subsequent levels of homeownership can create regulatory and behavioral opportunities for energy efficiency programs. New York City is largely a city of renters, which often leads to split incentives for energy efficiency between building owners and residents. At the regional scale, a suburban city government that does not have an interest in increasing public transportation options may make this a difficult strategy for reducing GHG emissions in the central city. National policies can provide opportunities for cities to push forward, such as Canada's new cap and trade policy or the Clean Power Plan in the United States. Indeed, city governments are quite strategic in their use of intergovernmental structures to further their own aims (Sellers 2002a).

Cities may choose to focus on those sectors that contribute the most to the overall level of GHG emissions. Every city has a unique mix of sectors and energy sources that are responsible for its GHG emissions. In some cities passenger vehicles represent a significant source of emissions, while building energy use represents a larger portion in others. For example, in Toronto, where population density is around 10,000 people per square mile, transportation is responsible for nearly half of the city's GHG emissions, while in New York City, where population density is 27,000 people per square mile, transportation registers at just 20 percent of the city's total. A city seeking to maximize its impact may choose to target the sectors that contribute the most GHG emissions. Alternatively, cities may choose to focus on the sectors in which they expect to have greater success. City governments look for opportunities to adopt popular policies and programs, capture co-benefits, or pick low-hanging fruit and ultimately target sectors where such opportunities lie (Betsill 2001). Again, such opportunities will be different for different cities depending on their larger context.

Choice of Governing Modes

Beyond a choice of sectors, city governments choose the mode of governing they use to target GHG emissions. Here I draw on Harriet Bulkeley's work on the modes governing cities use for climate change, their underlying logics, and relationship to a city's resources and capacities (Bulkeley and Kern 2006; Bulkeley 2013). In defining these modes of governing, Bulkeley (2013, 91–92) writes:

> In essence a mode of governing refers to a specific set of processes and techniques through which governing is pursued. Such processes can, in turn, be regarded as reliant on different forms of governing capacity and power, as encompassing tools, technologies and artefacts, and as based on different rationalities about the nature and purpose of governing climate change (Bulkeley, Watson, and Hudson 2007). . . . Although each mode is underpinned by different sets of processes, logics and techniques, they are not mutually exclusive; rather, municipalities tend to deploy a combination of these modes at any one time.

There are four such governing modes used by city governments to respond to climate change: self-governing, regulating, provisioning, and enabling. These modes represent a spectrum of intervention from direct to indirect, and each "relies on different sorts of policy instrument and intervention, as well as on the mobilization of different forms of resource and power" (Bulkeley 2013, 91). A choice of mode(s) is therefore a choice about the target constituencies and the use and necessity of particular sources of authority and resources.

Cities can *self-govern*, targeting the city government's own internal activities, such as energy used in municipally owned buildings and vehicles. The self-governing mode of governing climate change focuses on internal city operations—it is the city government changing its own operations and infrastructure in order to reduce GHG emissions. This can include energy efficiency measures for city-owned buildings, reducing or hybridizing the city government's vehicle fleet, installing solar panels or green roofs on city-owned buildings, and changing to low carbon energy sources for city operations. While city governments are often responsible for a small portion of the city's overall GHG emissions, this is where they hold the greatest level of formal authority. Nevertheless, it can still be challenging to coordinate the various departments and incentivize those in city government whose main responsibilities are not related to climate change mitigation to participate.

There are several reasons why city governments may choose the self-governing mode of governance: they avoid political debate by focusing on internal operations, they can usually be implemented fairly quickly, and they produce cost savings for the city (C. Gore, Robinson, and Stren 2009). Given these advantages, self-governing modes have been used in most cities (Bulkeley 2013; ARUP 2014). There are conditions under which cities may be more reliant on this mode for achieving their goals. When a city government faces significant obstacles from its broader policy environment (e.g., an unresponsive business community or hostile state government) or lacks political leadership to take on community-wide emissions, self-governing provides a means of moving forward on climate change mitigation. City governments without much experience with sustainability or energy-related programs can use the self-governing mode for experimenting and building capacity. Alternatively, a high-capacity city government will have the systems and structures in place internally that allow for corporate changes that significantly reduce GHG emissions, such as green purchasing, building retrofits, and the incorporation of new technologies like solar panels on city-owned buildings. Self-governing is also a tool for motivating or demonstrating to others (e.g., homeowners, the private sector) that the kinds of changes the city government is pursuing for the city are feasible or cost-effective. New York City has consistently set more aggressive goals for its own internal GHG emissions as a way of demonstrating its commitment and the feasibility of its goals. Using city government buildings as demonstration projects or test beds might motivate others to take action.

City governments can also use the *regulating* mode of governing to take direct action. This involves the use of traditional command and control approaches such as introducing new standards to municipal building codes, bans on certain fuels or practices, or land-use planning guidelines that dictate the technologies used and densities achieved in development activities. The regulating mode draws on the direct and formal powers of city governments; such actions are more transparent and the outcomes are easier to trace, but they may be more likely to attract opposition (Bulkeley 2013). For example, the city might choose to adopt bylaws that set higher energy efficiency standards in new buildings or introduce a tiered energy pricing scheme that charges large energy users more. Such measures have clearly defined criteria and expected outcomes and often last beyond a single administration, but they may not be popular with the development sector or certain populations in the city.

The ability of city governments to use the regulating mode of governing is necessarily dependent on the extent of their formal powers in relation to their GHG emissions. At a basic level, cities with greater authority over their GHG emissions are likely to rely more on the regulating mode of governance. This mode of governing comes with high political and administrative costs. Strong political leadership and high administrative capacity might therefore facilitate governing by regulating. Ambitious and proactive political leaders can help to change perceptions of the importance of climate change mitigation, the costs and benefits of aggressive climate change policies, and the feasibility of new approaches. Political leadership allows cities to prioritize climate change mitigation and target necessary collaborators. City governments can also find opportunities for regulating when the state government, or other local governments, are taking similar steps.

The third governing mode is *provisioning*, or governing through delivering new or altered services (Bulkeley 2013, 93). City governments often have purview over service delivery and infrastructure systems such as water, energy, transportation, and waste that can be changed in ways that significantly reduce the city's GHG emissions. For example, city governments can produce or purchase solar-generated electricity and use it instead of coal-generated electricity. City governments can introduce organic waste collection and recycling programs to reduce the amount of methane that is emitted from landfills. While city governments may not always control large transportation networks in their boundaries, such as highways and regional transit systems, they can provide climate-friendly transportation infrastructure by building bike lanes and electric vehicle charging stations.

Provisioning strategies such as these are able to target important sources of citywide GHG emissions, but they often involve significant capital investment in new infrastructure and concomitant behavior change on the part of those who use or help provide the city's services (e.g., people deciding to commute by bike). City governments with greater financial capacity, or access to resources, may be more likely to govern by provisioning. Alternatively, city governments better able to coordinate with other levels of government or service providers can more easily offer new or expanded services that reduce GHG emissions. City governments like Los Angeles, with a municipal energy utility, will also have an advantage, as they will be able to more directly target energy supplies and demand.

City governments can also choose to govern by *enabling*, which represents a more indirect mode of intervening. It relies less on a city's direct power and authority and more on alternative forms of power, such as the ability to influence, persuade, and educate (Bulkeley 2013). The enabling mode of governing climate change focuses on taking steps to facilitate or enhance the ability of other actors to reduce GHG emissions. Cities can invest in educational and awareness campaigns, provide technical or information resources, distribute grants, give planning guidance, or act as a "knowledge broker" for organizations or individuals working to reduce GHG emissions. This involves coordinating and facilitating partnerships by acting as a host or supportive venue, funding or participating in such partnerships, or removing legal or technical obstacles to such partnerships achieving their objectives. For example, decisions about using public transit can be shaped by the incentives they have for making that choice, such as the availability of a discounted bus pass. While it is increasingly common for city governments to govern using enabling approaches (ARUP 2014), it is a new type of strategy and one that may present a rather steep learning curve.

Governing by enabling requires that city governments provide others with the tools, incentives, and technologies they need to reduce GHG emissions. A high-capacity city government is better positioned to make such resources available and facilitate the necessary long-term monitoring and accountability. City governments without significant formal authority over their GHG emissions, or the political resources needed to intervene more directly, can rely on enabling modes of governing. Supporting and facilitating the actions of others may be the only option for a city like Toronto that does not control its energy supplies or have the ability to write its own energy code. Enabling is also a less heavy-handed approach and may therefore generate less political pushback from skeptics or the potentially regulated community.

Governing to reduce GHG emissions will require both direct and indirect modes of intervening. Each mode requires different sets of resources and capabilities, is likely to motivate and mobilize different participants and resources, and requires unique political and financial tradeoffs. Improving the energy efficiency of city government buildings in order to reduce GHG emissions (a direct, self-governing approach) only requires action on the part of the city government, requires technologies such as high efficiency light bulbs and HVAC systems, and is unlikely to require a complicated approval process. By contrast, incentivizing improved energy efficiency in commercial and industrial buildings (an indirect, enabling approach) requires action and expertise on the part of a number of individual building owners but likely does not require that city council pass additional bylaws. These are two ways of achieving similar outcomes (improving energy efficiency), but in using one or the other approach a city government will be achieving these outcomes through very different means and may see very different outcomes as a result.

A city's context is rarely static. New actors will arise, new incentives will come into play, and new opportunities for action will present themselves. City governments take stock of their progress along the way to see how well programs are working, to identify new or changing problem conditions, and to respond to shifting political and economic opportunities for progress. The strength and priorities of city leaders and private sector actors change. The formal authorities of city governments—such as their taxing and spending authorities or responsibility for certain types of services—can and do also change over time. The nature of the problem or governance task itself changes, particularly for a problem like climate change mitigation that requires long-term investment and attention. Political, economic, or cultural shifts can produce new commuting patterns in the city. A city's initial mission to reduce GHG emissions is likely to be reframed and adapted over time as political conditions and leadership shift. There will be opportunities for city governments to learn and improve, either based on their own experiences or the experiences of other cities.

City governments will adjust their agenda in response to changing contexts or as they learn about how well their efforts are working. A city may choose to prioritize their renewable energy goals when the state adopts a set of their own, or to frame their climate change programs as a job creation strategy when unemployment levels start to increase. If building owners are slow to take up required energy conservation standards, the city can decide to invest in capacity-building programs. City governments also adjust their approach as they learn more about what best practices in other cities look like. This is particularly likely to be true of cities that are members in international city networks such as the C40 or ICLEI, whose purpose is precisely to facilitate the sharing of lessons learned. Such feedback is essential for our understanding of the governance process in cities and how city governments effectively steer. In their analysis of the implementation of Empowerment Zone programs in the U.S., Rich and Stoker (2015) refer to this strategic adjustment as "program integrity," or the ability of a city to implement policy over time and make necessary adjustments to programs along the way.

While Bulkeley (2013, 98) notes that there are institutional, political, and sociotechnical drivers and barriers to city governments acting on climate change, she does not tie these drivers and barriers explicitly to variation in the choice of governing mode. Indeed, a number of scholars have called attention to the important role that a city government's larger technological and political–economic context will play as they attempt to shift to new patterns of energy and resource use (Hodson and Marvin 2010; Bulkeley 2013; Chester et al. 2014), but there remains little empirical work to demonstrate the relationship between the sources of urban GHG emissions and the governance strategies needed to change them (Monstadt 2009; Bulkeley 2010). In this framework, the choices city governments make about their modes of governing climate change mitigation are shaped by the dynamic set of obstacles and opportunities presented by their sources of GHG emissions, formal authorities, capacities, leadership strength, and larger policy context.

Part 2: Shared Governing Strategies That Mobilize Actors and Resources

City governments also make choices about the governing strategies they will use to develop and implement their climate change mitigation policies and programs. These are the strategies that activate governing capacities and allow city governments to launch new programs, pass new legislation, and engage new partners. Significant, transformative change in a city—such as those implicated by climate change mitigation—is not solely or primarily the product of progressive policies. The fabric of cities must be rewoven. As Stone argues, "the most feasible way to bring about change is not through pushing a different ideology; it is to alter cumulatively the organizational character of the 'playing field' . . . the battle is not an ideational one; it is one of using resources to construct a different network, to bring about a different civic and political milieu." (2015, 122).

Stoker (1989) similarly argues that "the essential task of implementation is to create a context in which participants are likely to cooperate to achieve policy goals despite the absence of a dominating authority." These prescriptions for change are distinct from discursive or behavioral approaches that suggest change comes from the introduction of new beliefs or access to political power. Rather, change is the result of deliberate strategies to restructure the political, institutional, and organizational networks that give rise to decision outcomes in the city. This work is the central task of city governments that have tasked themselves with the challenge of significantly reducing GHG emissions, where political leadership "is a creative exercise of political choice, involving the ability to craft arrangements through which resources can be mobilized, thus enabling a community to accomplish difficult and nonroutine goals" (Stone 1993, 1).

City governments must find ways to compel participation by individuals and organizations in the project of climate change mitigation. What new arrangements are needed for cities to make progress on reducing GHG emissions? What are the weaknesses of current urban governance arrangements that would prevent the assembly of the needed resources and motivation of participants to engage and change behavior?

Answers to these questions can be found in the three sources of uncertainty in urban climate mitigation discussed in chapter 1 that make reducing GHG emissions a "wicked" policy problem: institutional uncertainty, strategic uncertainty, and substantive uncertainty (Koppenjan and Klijn 2004).[2] Such uncertainties make it difficult for city governments to pursue and promote collective action on climate change. Therefore, effective governing strategies help to mitigate, if not resolve, these sources of uncertainty in urban climate change governance. Specifically, building new institutions helps to resolve institutional uncertainty, building new coalitions helps to resolve strategic uncertainty, and building new capacities helps to resolve substantive uncertainty (figure 2.2). In removing or reducing

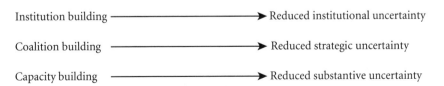

FIGURE 2.2 Governing strategies for reducing uncertainties in climate change mitigation

these sources of uncertainty, these governing strategies create conditions more conducive to collective action on climate change mitigation, mobilizing the resources and actors needed for cities to travel their chosen route to significant GHG emission reductions. They "alter cumulatively the organizational character of the 'playing field'" (Stone 2015, p. 122) in ways that are directly relevant to the challenge of reducing GHG emissions.

Building New Institutions

Institutional uncertainty arises from the fact that urban climate change mitigation cuts across existing sectors and administrative demarcations. People in different departments and different levels of government will be guided by the "tasks, opinions, rules and language of their own organization, administrative level, and network," which can lead to clashes between divergent institutional norms (Koppenjan and Klijn 2004). New institutional structures are likely to be required to authorize new projects, gain necessary approvals, monitor progress, and provide enforcement. Ensuring that institutions reflect new commitments and cultures is foundational to implementing innovative policy (Steelman 2010). As a cross-cutting and long-term policy issue, there is uncertainty about where to "house" climate change mitigation and whether there is sufficient institutional capacity to monitor and protect mitigation efforts over the short and long terms. Governing to reduce urban GHG emissions will require that city governments build the institutions necessary to assemble and coordinate participants and resources.

Integrating climate change programs into existing institutional structures is a major challenge for city governments (Aylett 2014), but it is often a first step in the implementation process. Anguelovski and Carmin (2011, 2) find that "as local governments devise climate policies, they tend to formalize and institutionalize their work in order to facilitate implementation and strengthen the legitimacy, coordination, and support for such policies across sectors and departments." Choices must be made about where to locate programs and how to staff them (e.g., in the mayor's office or in departments), whether to mainstream climate change mitigation throughout existing offices and programs or to create a central hub for climate change activities, and how to fund the necessary programs (Hawkins et al. 2015). Portney (2013, 223) argues that to be successful, city governments must find ways to "tie the specific carbon reduction strategies to specific administrative units, agencies, or departments, and then make sure that those units have the authority and funding to achieve the reductions." This creates a useful cohesiveness to the implementation process (Hjern 1982; Rich and Stoker 2015).

A key institutional challenge for cities has been, and will continue to be, ensuring that their efforts to mitigate climate change persist through changes in political leadership and staff turnover. Reducing urban GHG emissions is a long-term endeavor, requiring ongoing feedback and learning processes to govern effectively. New ideas, new financial tools, and new standards of practice are necessary to reduce GHG emissions in cities. Fostering these types of innovations, and scaling them up, is necessary for cities pursuing deep GHG emission reductions. Innovations take time to develop and can come into conflict with existing mandates and decision-making procedures and generate significant transaction costs.

Institution building allows climate change work to continue through political turmoil and shifts. Many innovations, such as building a large supply of solar energy, require years if not decades to produce measurable GHG emission reductions, and the institutionalization of climate change mitigation will help to ensure that climate change policies and programs persist through administrative changes. Such institutionalization can also lead to new ideas as agencies and stakeholders seek to meet new expectations or leverage new incentives. Climate change mitigation needs to be embedded into the formal and informal institutions that guide decision making in cities to ensure continuity and broad adoption, contributing to learning and mitigating the effects of political transitions.

Building New Coalitions

Strategic uncertainty in urban climate change mitigation arises from the wide range of actors implicated in mitigation, each with a unique set of preferences and resources. For city governments to move forward they need buy-in from, and coordination with, the private sector, residents, decision makers, and key stakeholder groups. It can be difficult to predict or even influence any particular actor's or group's choices about energy use or modes of transportation and how these choices will interact to influence GHG emission patterns and decision-making processes. One antidote to this uncertainty is to broaden and strengthen the coalition supporting and promoting climate change mitigation efforts in the city. City governments can take steps to build support for their plans and foster a network of advocates and supporters rather than sources of resistance or uncertainty. This is very much a relational model of urban governance in that the capacity and ability of city governments to govern is activated and strengthened by the partnerships and relationships they are able to generate in the process.

Coalition building allows city governments to build support for plans and policy programs, foster a network of advocates, and increase their access to expertise. This model of governance is not unique to cities; environmental governance more broadly has increasingly turned to more collaborative approaches. As policy problems have become increasingly complex and the context of governance has changed the role that government plays in solving them, collaboration with diverse sets of stakeholders has become necessary to make progress (Emerson and Nabatchi 2015). Governments can play a variety of roles in the subsequent partnerships (Koontz et al. 2004, 174). Coalition building for urban climate governance includes but extends beyond collaborative governance models. City governments use a variety of tools and strategies to generate political support and coordinate critical actors; coalition building is an active process for city governments. As Kristin Good points out, developing a broad coalition "depends on the capacity and willingness of municipal officials and civil servants to serve as facilitators or 'bridges' and on the ability of these agents to *recognize opportunities for strategic cooperation*" (2009, 33).

Potential collaborators or coalition members are motivated by material incentives, purpose, and "small opportunities" (Stone 2005). Framing the benefits of GHG emission reductions measures broadly—for example, highlighting their job creation potential or expected cost savings for residents—makes these opportunities apparent, which can be a tool for coalition building. Indeed, effective framing by city governments has consistently been found to be an important factor in determining their success in governing GHG emissions (Betsill 2001; Bulkeley 2010; Ryan 2015; Laurian, Walker, and Crawford 2016). In *The New Localism*, Katz and Nowak describe the strategies of "vanguard cities . . . catalyzing growth through forms of governance that align the distinctive perspectives of government, business, philanthropy, and the broader community" (Katz and Nowak 2017, 3).

City governments can choose to frame their climate change programs as a job creation strategy, as a way to improve residents' health, or as a means for meeting other priorities such as reducing traffic congestion or saving money on energy bills. The city of Toronto's plan to reduce GHG emissions frames the programs as a way to improve local air quality and achieve a state of "energy sustainability" (Bierbaum et al. 2013), while Los Angeles highlights the opportunity to "promote cutting-edge green technology that can be marketed around the world." These framing decisions can help generate community and interest group buy-in and provide a way for decision makers to meet broader objectives. The framing of climate change programs might also change over time as a way of maintaining a coalition for action (Stone 2015). In this sense the need for coalition building can also shift policy aims and purpose as new interests are brought on board.

I also include within coalition-building strategies for generating support at the ballot box for candidates and legislation. For example, a city may be interested in introducing new building energy standards for commercial buildings. Broad political support must be generated to pass such a bylaw, including from those without a direct interest in the environmental or climate benefits of such a measure. City governments can frame the new energy use standards in a way that emphasizes benefits with broader appeal, such as the jobs created from new technology or the competitive advantage for the city of having cutting-edge standards in place. They can also take measures to engage stakeholders in the decision-making process, which can generate buy-in and understanding from groups that might normally oppose stricter energy use standards.

Coalition building takes place through engagement and active coordination. City governments make important decisions about the implementation process—who is involved, at what point in time, through what means, and with what level of interest. Research has consistently shown that when city governments engage residents and stakeholders in the implementation process they are likely to have greater success with environmental programs (Portney 2013; Ryan 2015). Participation and engagement facilitates greater transparency and trust in government-led initiatives and can generate important sources of new information for administrators tasked with program delivery. This can take the form of stakeholder advisory groups such as New York City's Green Codes Task Force, formal and informal consultations, public meetings, and other means of actively seeking to negotiate and link up agendas. Building and broadening a climate change mitigation coalition in cities will therefore be a long-term and ongoing process of engagement, coordination, and communication.

Building New Capacities

Substantive uncertainty arises from the availability and usability of information, as well as the interpretation and meaning of new information. Reducing GHG emissions is a new aim for city governments; most places have very little previous experience to draw upon. City governments are therefore likely to face uncertainty about how mitigation actions relate to outcomes (both in the short and long terms), how to best design or administer new programs, or even basic trends and features of their GHG emissions profile. Key stakeholders, such as private building owners or energy utilities, similarly lack experience and capacity in adopting the kinds of measures necessary to reduce GHG emissions. Reducing or resolving this uncertainty therefore requires that city governments recognize, commit to, and invest in building new capacities both internally and in key stakeholders and partners. Capacities of various kinds are consistently associated with greater progress on climate change mitigation (Zahran et al. 2008; Krause 2011b; Sharp,

Daley, and Lynch 2011). City governments actively work to generate and build new capacities as they pursue GHG emission reductions, which further serves to mobilize resources and actors.

Governmental capacity is necessary to generate clear directives and appropriate targets, enforce policy through monitoring and accountability mechanisms, and manage the implementation process internally through the coordination of personnel and resources (Mazmanian and Sabatier 1981). The human resource needs of climate change mitigation can be significant. Developing programs, managing funds, and interpreting information require that competent people be tasked with completing them. North American cities are most likely to report having only one staff member working on climate change (Aylett 2014). While climate change is often a mayoral initiative, their offices are often understaffed. Technical resources (such as an energy use data collection system or GHG emissions inventory) provide the information necessary for setting priorities, identifying appropriate technologies, and monitoring progress.

Tracking GHG emissions often requires collecting new data or compiling existing data in very new ways. City governments may not have data sharing agreements with their energy utility, or they may even find data sharing between internal departments such as sanitation and parks to be difficult and cumbersome. There may also be political sensitivities associated with GHG emissions such that city governments, or individual departments within cities, are reluctant to share detailed information about where energy and fuels are most heavily consumed in the city. When Toronto was unable to acquire updated transportation emissions data from the city's Department of Transportation, it worked with the province and academics to build a new dataset tracking vehicle miles traveled in the city. City governments may not have the technical expertise in-house to convert energy use and fuel consumption data to GHG equivalents. To effectively govern climate change mitigation, city governments will need to build new capacities internally among staff and technical systems.

Participating and engaging in global city networks is a strategy city governments are increasingly using to build new capacities and reduce substantive uncertainty. Networks like the C40 or ICLEI's Green Climate Cities (GCC) can expose city governments to new norms, new ideas, and new sets of technical resources that can shape the decisions city governments make about their own approach to reducing GHG emissions and the outcomes they are likely to achieve as a result. The C40 network in particular is increasingly focused on establishing best practices and facilitating city-to-city learning (Lee 2012; ARUP 2015). The GCC will waive member fees in exchange for allowing an ICLEI staff member to sit in the city government and assist in program development.[3] Other initiatives, like the Rockefeller 100 Resilient Cities, actually fund municipal staff members to take up work to respond to climate change. These networks have recognized the importance of capacity building for urban climate change governance and are taking steps to help cities strengthen their own capacities.

Mobilizing resources and participants requires investing in the capacities of stakeholders. Reducing GHG emissions in cities often demands individuals and organizations make new choices about how they invest, what they buy, how they behave, and what they build. Changing the incentive structures around such decisions—for example, by requiring tracking and monitoring, providing rebates, or introducing new financial tools—is an obvious and attractive way to work toward different outcomes. But in some cases this may not be enough. New standards and incentives also require knowledge and resources from stakeholders and must be complemented with capacity building if cities hope to see real change. This includes fostering greater awareness, education, and access to needed technical resources for key stakeholders. Significantly reducing GHG emissions requires an investment in stakeholder capacities in order for them to meet new requirements and standards.

While each city faces a unique landscape of emissions sources and political and institutional leverage points, the underlying uncertainties and governance challenges of climate change mitigation are largely the same. The requirements for governing climate change mitigation are constant: city governments must alter cumulatively the landscape that underpins their GHG emissions. This requires building the new institutions, coalitions, and capacities that activate and mobilize the resources and actors needed to pursue a mitigation plan. Therefore, while the sectors and governing modes city governments choose for reducing GHG emissions are likely to vary significantly, the centrality and use of these three governing strategies will underpin climate change mitigation in all cities.

Part 3: Multidimensional Evaluation of Impacts

Reducing urban GHG emissions requires reshaping cities in ways that accomplish shared climate change mitigation goals. Evaluating systematically these outcomes, "how systems of governance deliver policy-relevant impacts" (O'Toole 2000, 276), is a necessary component of urban climate governance analysis. Are the efforts of city governments working? What and who is being affected? If we are to understand how city governments govern climate change, we need to track closely where they are taking their cities.

The primary aim of city governments working toward climate change mitigation is to reduce GHG emissions, and measuring a *change in GHG emissions* is therefore an intuitive choice for assessing the progress cities are making. Mitigating climate change requires measurable and dramatic reductions in GHG emissions, and the contribution of city governments to this environmental outcome would therefore most directly be measured as changes in GHG emissions they can claim to have made. In this sense, a city government that has reduced its city's GHG emissions by 10 percent over a ten-year time frame would be making less progress than a city government that has reduced its city's GHG emissions by 25 percent over the same time period.

Changes in GHG emissions are also a politically useful metric: this is the claim decision makers would like to be able to make publicly after investing city resources into climate change mitigation programs. For example, if a city government pledges to reduce GHG emissions 20 percent below 1990 levels by 2020, it is important to know the depth of GHG emissions reductions that have been achieved as of 2016. For public accountability, this is in many ways the litmus test for city governments and climate change mitigation.

There are drawbacks to relying exclusively on GHG emissions in tracking a city's destination. Urban GHG emissions are very difficult for cities and scholars to actively track and monitor. Many cities, even those that have been engaged with climate change mitigation for a decade or more, have only completed one or two GHG emission inventories. Completing a GHG emission inventory can be very resource intensive for a city government (D. Pitt and Randolph 2009). If a city government has not completed a GHG emission inventory and is not in a position to share the necessary data, it can be nearly impossible for scholars to complete an accurate inventory independently.[4]

It can also be extremely challenging to attribute changes in a city's GHG emissions to a given project, program, or policy. A city's GHG emissions are the product of a complex web of factors including the weather (e.g., a cold winter), broader economic conditions, policy choices made by other governments, individual behavioral choices, and the actions of city governments. Even identifying the potential GHG emission savings that are possible from a particular program or policy can be technically challenging, and attributing a 5 percent decrease (or increase) in the city's GHG emissions to the implementation of a particular city government program (or lack thereof) can be quite difficult. Relying on GHG emissions as a single metric of effectiveness could therefore provide an overly optimistic or pessimistic view of how effective the city government has been in implementing its climate change mitigation plan. Using GHG emissions as a single metric of effectiveness also obfuscates both the potential broader implications of city government actions and the mechanisms by which the implementation of a climate change mitigation plan leads to GHG emission reductions.

Evaluating a city's progress, and understanding its destination, must extend beyond just trends in GHG emissions. Given the fundamental changes to a city that are necessary to reduce GHG emissions, evaluating the destination should also include the broader consequences of reorienting urban emission systems, or what has been called the city's "catalytic impact" (Bulkeley et al. 2014). Responding to climate change requires radical and sustained changes to urban systems. Such broader system shifts can be observed in several ways.

City governments can succeed in changing behaviors. Behavior change is a useful metric for evaluating the consequences of city government action as it is "closer in the causal chain to institutions than is environmental quality. This means there are fewer—even if not few—alternative explanations of why behavior changed than of why environmental quality [e.g., GHG emissions] changed" (Mitchell 2008, 84). Behavior change can come in the form of increased resource allocation to climate change programs (Wheeler 2008), increased inter-agency collaboration, and energy conservation in municipal activities. Other behavior changes will need to take place outside of city government, such as changes in household commuting patterns, compliance with voluntary or regulatory energy efficiency programs, and installing solar panels on commercial, institutional, and residential buildings. Metrics such as increase in public transportation use, number of companies reporting their energy use, and change in the proportion of the city's energy supply coming from renewable sources can be very useful metrics of implementation effectiveness.

It is also important to understand the indirect effects city government might have on other parts of the city. For example, if a city is partnering with large commercial building owners to promote energy efficiency, smaller operations may see their successes and adopt efficiency measures of their own. Local environmental organizations may feel empowered to develop programs of their own that build on and expand the successes of the city-led efforts. Conversely, a failure to include marginalized groups or certain stakeholders in the city in climate change mitigation programs can lead to further marginalization and reinforcement of existing power structures (Rosan 2012).

The efforts of city governments may compel state, provincial, or federal governments to follow their lead.[5] Selin and VanDeveer (2007) propose this could take place through one of four pathways: the strategic use of demonstration effects, market expansion and pricing, policy diffusion and learning, and norm creation and promulgation. Indeed, in many city climate change plans there is the explicit hope that other levels of government will follow the city's lead and support its efforts to reduce GHG emissions. Cities can also have an indirect impact through city-to-city transmission of ideas, experience, and strategies. This could take place through city climate change networks such as the GCC or C40, or through other city networks such as the U.S. Sustainability Directors Network.

While indirect broader system shifts are not likely to directly result in GHG emission reductions for the city, they are likely to contribute to the larger task of addressing climate change. A city that is very effective at initiating such shifts should be considered a positive contributor to climate change solutions even if their direct GHG emission reductions are lackluster.

Applying the Framework

Reducing GHG emissions and contributing to climate change mitigation is not a product of certain conditions being met (e.g., mayoral leadership or state-level incentives) but rather of deliberate actions and choices made by city governments to make use of the opportunities that are present and build the structures needed for collective work. The ability of city governments to contribute to climate change mitigation is contingent on their ability to mobilize participants and coordinate resources, governing tasks that city governments are uniquely positioned to take on and that do not always require formal authority or significant financial resources. Given the multiple modes of governing available to city governments, and a shared set of governing strategies, cities in very different contexts can contribute to reducing GHG emissions.

There is no way to predict which cities will be more effective based solely on their characteristics or context (e.g., large cities, liberal cities, wealthy cities). Rather, we should expect there to be multiple paths to similar impacts. Even when the agenda being pursued by cities is the same, the type and quantity of participants and resources that must be mobilized to realize this agenda are place-specific. The choices city governments make about how to govern will reflect their particular opportunities for channeling collective action. City governments can make progress, even when they lack formal authority or face resource constraints, by making different choices about how to mobilize needed resources and participants. The key for city governments will be developing governing strategies that mobilize the participants and resources needed at a particular place and time. We can expect to observe multiple combinations of direct and indirect interventions that produce similar impacts as long as they are supported by the three shared governing strategies. This framework is therefore not specific to a certain type of city but provides a means of evaluating and interpreting climate change governance in any city that has committed to working toward a GHG emissions reduction goal.

Repowering cities requires evaluating and harnessing the tools and strategies available to city governments to lead and govern their cities. This framework provides analytical leverage for tracking how city governments are activated and the choices they make when navigating their broader context. As such, the framework also helps to identify why city governments might *fail* to govern or when policies and programs fail to produce their intended results. There are several new governing tasks within the landscape of the urban inversion, including addressing inequality, immigration surges, or infrastructure deficits. The fragmentation of power and resources in urban governance is particularly salient in such policy environments (Sapotichne and Jones 2011; Bramwell and Pierre 2013; Stone 2015); this is where the gap between the challenges facing cities and the formal capabilities of cities is the most "overwhelming" (Stone 1989). This framework is applicable to these new, complex, and uncertain policy aims as it provides a way of distinguishing between the policy choices city governments make and the governing strategies they employ, and the local and catalytic impacts of urban governance.

This analysis of repowering cities is centered on the challenge of reconfiguring and reorienting the infrastructures, behaviors, institutions, and economies that power our cities in order to reduce their GHG emissions and contributions to global climate change. I use this framework to analyze and interpret the experiences with climate change mitigation of three case study cities: New York City, Los Angeles, and Toronto. Each committed in 2007 to significantly reducing their GHG emissions (table 2.2).

As relatively early actors, these three cities have more experience with governing climate change mitigation than most and provide an opportunity to track

TABLE 2.2 Summary of the climate change commitments from New York City, Los Angeles, and Toronto in 2007

CITY	NAME OF PLAN	RELEASE DATE	GHG REDUCTION TARGET
New York City	PlaNYC: A Greener, Greater New York	April 2007	30% below 2005 levels by 2030
Los Angeles	Green LA: An Action Plan to Lead the Nation in Fighting Global Warming	May 2007	35% below 1990 levels by 2030
Toronto	Change is in the Air: Climate Change, Clean Air and Sustainable Energy Action Plan	June 2007	6% below 1990 levels by 2012 30% below 1990 levels by 2020 80% below 1990 levels by 2050

decision making and progress over time. They also have useful similarities and differences that provide an opportunity to evaluate the three central ideas developed in this chapter: that city governments tailor their choices to their context, that there are three governing strategies that mobilize resources and actors, and the impacts of this work can be both GHG emission reductions and broader, catalytic impacts beyond the city. I trace carefully the experiences of these cities with governing climate change mitigation in order to understand what progress they have made and what potential there is for greater transformation.

3

MADE TO MEASURE

Tracing Unique Climate Policy Agendas
in New York City, Los Angeles, and Toronto

Governing climate change mitigation requires that city governments navigate a unique political, institutional, economic, and technical context. Each city has a distinctive mix of GHG emission sources depending on its morphology, economy, and social dynamics. City governments also bring with them different levels of authority, leadership, and capacity. Their broader policy and political contexts will be more or less conducive to reducing GHG emissions. City governments in a range of such contexts have made commitments to reduce their GHG emissions; as they work to meet these goals, their climate change policy agenda will be shaped by the context in which they govern.

City governments can make choices about the sources of GHG emissions they target, prioritizing those that contribute the most to the city's overall emissions or that present an easy win or "low-hanging fruit." There are also various governing modes city governments can use to enact and facilitate change. These governing modes involve more or less direct intervention and present means for cities to act even when they lack formal authority or face a challenging political environment. The potential to tailor a city government's policy agenda to its particular context is important as it can allow many different types of cities in different circumstances to address climate change in a meaningful way.

In this chapter I trace the climate change mitigation policy agendas of New York City, Los Angeles, and Toronto, focusing on the period since 2007, and specifically the tools with which they sought to reduce GHG emissions. I first describe each city's unique context, emphasizing initial conditions in 2007, when they each made a formal pledge to reduce GHG emissions, but also the important changes that have taken place since then. I then unpack the primary features of each city's mitigation policy agendas in terms of the sources of GHG emissions they have targeted and the governing modes they have employed to pursue change. I use as indicators the policy priorities of mayoral administrations and city councils, subsequent staff reports and planning documents, the development of new programs and initiatives, and financial investments. Each helps to reveal the paths these cities have chosen for reducing GHG emissions.

The climate change mitigation policy agendas developed by the three cities do reflect their particular context and the process of learning and strategic adjustment over time. Each city has a very unique set of programs and policies in place to target GHG emissions, from energy use benchmarking to installing solar panels to incentivizing alternative modes of transportation. In each case this mix initially reflected the city's strengths and opportunities. Over time, the cities have expanded and diversified their agendas in response to changing conditions and new information. In some cases the cities have had to readjust when certain approaches failed or lost political support. In other cases the cities have sought to take advantage of new opportunities or diversified their agendas in an effort to meet more ambitious goals.

Ultimately, the experiences of New York City, Los Angeles, and Toronto reveal the unique and varied policy agendas city governments develop for reducing GHG emissions and the twists and turns these agendas take over time as conditions change and decision makers learn. Their experiences highlight the variation in both the task of urban climate change mitigation and the choices available to city governments for governing climate change in their city.

The Context of Climate Change Mitigation in the Three Cities

The task of reducing GHG emissions in New York City, Los Angeles, and Toronto reflects the varied technical, political, economic, and social characteristics of the three cities (table 3.1). The relative GHG emission contributions of energy use, transportation, and waste vary considerably: from New York City where building energy use contributes more than 70 percent of the city's total, to Toronto where it contributes just over half. Electricity in Los Angeles is provided by the largest municipally owned utility in North America—the Los Angeles Department of Water and Power—while in New York City and Toronto electricity is provided by public and private entities largely regulated at the state or provincial level. These contours will ultimately shape the agendas the three cities develop to meet their ambitious goals for climate change mitigation.

TABLE 3.1 Summary of three case study city terrains, highlighting the key features of each element

	EMISSIONS SOURCES (MOST RECENT INVENTORY)	AUTHORITY	CAPACITY	POLITICAL LEADERSHIP	BROADER POLICY CONTEXT
New York City	80% energy use 20% transportation	*High:* City regulatory powers	*Medium:* Large, well-funded government; Little experience	*High:* Signature mayoral project	*Low:* State and federal government support tepid
Los Angeles	60% energy use 34% transportation	*Medium:* City provisioning powers	*Medium:* Moderate financial resources; Some experience	*Medium:* Mayoral project but lacks support from council	*High:* State support is uniquely high, federal support tepid
Toronto	55% energy use 41% transportation	*Low:* Few sources of formal authority	*Medium:* Limited financial resources; Dedicated, experienced staff	*Low:* Often lacks clear elected champion	*Medium:* Provincial support is high, federal support mixed

New York City

New York City has a GHG emissions profile that is distinct from other North American cities. In 2005, nearly 80 percent of New York City's 58.3 million metric tons of GHG emissions came from energy used in buildings.[1] This is due to the fact that New York City is the densest city in the U.S., at 27,000 people per square mile, and relatively well served by public transportation systems. The majority of New Yorkers (65 percent) walks or takes transit to get to work (City of New York 2007b). While for many cities transportation can account for as much as 50 percent of GHG emissions, in New York City it is only responsible for around 20 percent of the total.

Reducing energy used in New York City's buildings requires the retrofitting of existing buildings, as 80 percent of the city's 2030 building stock already exists.[2] Half of New York City's energy-related GHG emissions are produced by electricity generation and half are produced by the direct use of heating and cooling systems in buildings. A demand-side approach to reducing GHG emissions is a challenging prospect for any city, as building stock turns over very slowly, perhaps every 30 years, and retrofitting buildings is technically and financially challenging.[3] One advantage New York City has is that a great deal of its energy use takes place in very large buildings. Most New Yorkers live in multifamily housing, and commercial, industrial, and institutional buildings account for 60 percent of energy-related emissions. The result is that only 4 percent of the city's buildings contain 50 percent of the city's built area. This is useful for policymaking as it provides the potential for large impact from retrofitting a relatively small proportion of the city's buildings.

Like most cities, New York City has the ability to develop standards for city-owned buildings, which account for around 4.5 percent of the city's GHG emissions, and vehicle fleets. New York City is able to adopt its own energy codes and standards for private building construction and retrofitting, permitted they are more stringent than those developed by the state government.

Overall, despite being broadly considered a strong city government, New York City has very little formal authority over its energy supplies and transportation systems. There are eight distinct federal, state, and local entities that have some responsibility for energy planning for New York City. The State of New York deregulated the energy sector in 1998 to foster competition, and the city's private utility, Consolidated Edison (ConEdison), was forced to sell a good portion of its assets.[4] Most electricity (75 percent) for the city comes from in-city power plants and is then supplemented with short- and long-term contracts with producers in upstate New York using natural gas, coal, oil, and nuclear power (New York City Energy Policy Task Force 2004). ConEdison provides a mix of natural gas and heating oils for heating and cooling New York City's buildings.

Transportation planning and financing for New York City takes place primarily through the regional Metropolitan Transportation Authority (MTA), "a quasi-autonomous public authority" (Chronopoulos 2012, 199). Ultimately, the MTA is responsible to the governor of New York, who is able to appoint and remove the seventeen board members. Vehicle emission standards are also largely out of the city's control, though the city can set emission standards for city-owned vehicles.[5] New York City has a Department of Transportation that engages in planning and oversees transportation policies and is primarily charged with the maintenance of the city's streets, sidewalks, highways, and bridges (S. Schwartz et al. 2009).

A strength New York City brings to reducing GHG emissions is the political leadership and commitment from both the mayor's office and city council. With the release of *PlaNYC* in 2007 by Michael Bloomberg, the mayor of New York City was, for the first time, fully invested in an environmental agenda for the city and pursuing climate change mitigation as part of the city's long-term planning strategy.[6] Under Mayor Bloomberg, the city was a founding member of the C40, and Bloomberg served as its chair from 2010 to 2013. The city hosted the C40 in 2007, the United Nations Climate Summit in 2014, and the World Cities Summit Mayors Forum in 2015. This political leadership carries special weight in New York City because it is a strong city government with a strong mayor system. The mayor largely sets the city's agenda.[7] Marcia Bystryn, president of the New York

League of Conservation Voters, is quoted as saying, "In city government, if you're not convinced the mayor is behind the initiative, you probably don't spend a lot of time on it, because you know it's just never going to have any real traction" (Bagley and Gallucci 2013).

New York City is also very large—the largest city in the U.S., with 8.5 million people—and brings important sources of capacity to the challenge of reducing GHG emissions. The city was in very good financial shape in 2007.[8] The city has multiple own-source revenue streams—including local property, sales, and income taxes—totaling more than $7,000 per capita, and a total operating budget of over $60 billion (City of New York 2007a).[9] As Bloomberg pointed out when announcing *PlaNYC*, "our economy is humming, our fiscal house is in order and our near-term horizon looks bright. If we don't act now, when?"

Administratively, New York City is also the largest city government in the U.S., with over 250,000 employees and major purchasing power in the energy marketplace.[10] Given the lack of executive leadership on climate change prior to 2007, the city did not have much technical expertise or planning experience relevant to tracking GHG emissions or developing policies designed specifically to reduce GHG emissions.

The city's broader policy and political context provides a mixed set of opportunities and obstacles. Internally, leadership on important issues is something New Yorkers are relatively familiar and comfortable with, and are even likely to expect of their city government. Decision makers in City Hall view New York City as an innovator and a city that is able to get things done even when others can't.[11] Due to its size and economic productivity, the city also occupies an influential position in the region and state. Decision makers often even feel pressure to lead and to maintain a leadership position over time, though without jeopardizing the city's global position as competitive and economically productive.[12] Influential political interests are relatively well organized and centralized within the city, through such groups as the Real Estate Board of New York and various philanthropic and environmental organizations.[13]

Until recently the State of New York has had quite modest climate change goals and no requirements of city governments. New York has been a member of the Regional Greenhouse Gas Initiative (RGGI) since 2005, one of the country's most successful and durable carbon emissions cap and trade networks (Rabe 2016). The state did not meet its renewable portfolio standard (RPS) goal of increasing renewables from 19 percent to 25 percent by 2013. Federally, the U.S. has consistently failed to pass legislation through Congress that would limit GHG emissions. The Obama administration took steps to strengthen federal standards and regulate GHG emissions, capitalizing on what executive authority is available for reducing GHG emissions and helping to shift public discourse on the issue. For example, the administration's 2009 Recovery Act stimulus included funding for energy efficiency (Energy Efficiency and Conservation Block Grants) and renewable energy.

In response to growing concern about dependence on foreign oil, the U.S. government passed the Energy Independence and Security Act in 2007, which set standards for renewable fuels and energy conservation technologies and improved fuel efficiency standards for automobiles to thirty-five miles per gallon. The U.S. Environmental Protection Agency began to develop regulations for GHG emissions, and in 2015 it released the Clean Power Plan that sets GHG emission targets and standards for electricity production that states must meet. Implementation has been stayed by a challenge from twenty-seven states in the Supreme Court and, following Donald Trump's election in 2016 and the partisan and polarized nature of climate change policy in the U.S. (McCright and Dunlap 2011), the likelihood of comprehensive federal legislation remains low. Rather, New York City has occasionally sought small grants from various federal agencies to support specific projects.

Los Angeles

Tracking GHG emissions for the city of Los Angeles is challenging. The city has released just two emission inventories, in 2007 and 2016, and used two different methods. As a result, they paint very different pictures of the city's emissions profile. In 2007, Los Angeles reported its total emissions for 2004 as 51 million metric tons, down from a peak of 54.1 in 1990 (City of Los Angeles 2007). While this is slightly less than New York City, it positions Los Angeles as a much more emissions *intensive* city, responsible for over 13 tons per person compared to New York City's 7. At the time, the city attributed nearly half of citywide emissions to transportation, primarily cars and trucks (City of Los Angeles 2007).

At just 8,400 people per square mile, Los Angeles has around one-third of New York City's density. This dispersed urban form and historic underinvestment in public transportation affects mobility patterns in the city: more than 75 percent of Los Angeles residents drive to work, most of them alone, and just 15 percent of Angelenos take public transportation, bike, or walk to work (U.S. Census Bureau 2013). Unsurprisingly, congestion in Los Angeles is the worst in the country, and commuters spend an average of eighty-one hours each year in traffic (Nelson 2016).

The city also reported in 2007 that municipal operations accounted for one-third of the city's emissions, likely due to the fact that the city is powered by a municipally owned utility, the Los Angeles Department of Water and Power (LADWP). Using figures reported by the LADWP to the California Climate Action Registry, the city's energy use contributed 13.2 million metric tons of GHG emissions, or around 25 percent of the city's reported GHG emissions for 2005.

In addition to a municipal energy utility, Los Angeles owns and operates the country's busiest port, governed by the Los Angeles Port Authority, a proprietary department of the city. The Port of Los Angeles is responsible for more than 40 percent of U.S. imports, and in 2015 over 150,000 trucks came in and out of the port (Port of Los Angeles 2016). Similarly, the city owns and operates the Los Angeles International Airport. These special features of Los Angeles can make GHG emissions accounting challenging and difficult to compare with those from cities without such assets.

In 2016 the city released its first updated emission inventory, this time using the new Global Protocol for Community-Scale GHG Emissions Inventory. The new inventory includes estimates for 1990 and 2013 and paints a very different picture of the city's emission sources. Here, the city reports producing 36.2 million metric tons of GHG emissions in 1990, or around 10 tons per capita. In both 1990 and 2013, transportation emissions are estimated to account for only around one-third of the city's total, with energy use responsible for the bulk of the remaining emissions (City of Los Angeles 2016).

The change in reported emissions is largely driven by a bottom-up accounting of transportation emissions that greatly reduced their contribution to the city's total.[14] This means that rather than disaggregating state or county-level transportation emissions on a per capita basis, the city devised its own dataset of transportation patterns within the city and the emissions associated with them. Ultimately, the city has been operating without a clear understanding of its sources of GHG emissions. When they began their work in 2007 they understood transportation to be the largest contributor of GHG emissions, followed by energy use. These numbers reversed in 2016 with the release of the new inventory.

Unlike New York City, the City of Los Angeles is able to play a more direct role in shaping both the city's energy demand and energy supplies. The LADWP operates as a proprietary city department, which means that it manages and controls its own assets and has access to tax-free bonds. The department's revenues are generated entirely from user fees from water and energy customers; it does not receive general tax revenues from the city. Energy and water rates must be approved by the city council, which is a source of political conflict for the city.

A five-member Board of Water and Energy Commissioners, appointed by the mayor of Los Angeles and confirmed by the city council, governs the LADWP. The relationship between the city and the LADWP has shifted over the years and at times can be quite contentious (Hughes, Pincetl, and Boone 2013). While the California Public Utilities Commission regulates the primarily private energy utilities in California, the LADWP has a tangential relationship to state regulation: some standards apply but the state will typically provide greater flexibility and fewer direct accountability mechanisms. The LADWP often uses state standards as a metric or benchmark for its own targets and actions.

The LADWP is responsible for the city's energy supplies, long-term energy planning, and energy supply contracts. In 2007, around half of the city's energy was purchased from coal-burning power plants in Utah and Arizona, 25 percent came from natural gas power plants in the Los Angeles region, and around 8 percent came from renewable sources like solar and wind (City of Los Angeles 2007). The privately held Southern California Gas Company (SoCal Gas) provides natural gas to the city.

Transportation governance in Los Angeles is similar to New York City's, but the governance structure of the regional transportation body provides the city more options for directly influencing planning and investment. The Los Angeles County Metropolitan Transportation Authority (LA Metro) is the regional agency responsible for planning and financing transportation projects for the county. It was established by state legislation, and has thirteen board members who vote on transit priorities and funding. The mayor of Los Angeles is one board member and is able to appoint three others. This can translate to significant influence on the board, as only three more votes are needed for a majority.[15] New York City does not have such influence on regional transportation planning. The Los Angeles Department of Transportation, a city department, plays a role in planning for bike lanes and local traffic management, including the potential for dedicated bus lanes.

The mayor's office and a small handful of city councilors have provided political leadership on climate change in Los Angeles. Political leadership on climate change has not always been a unified front in Los Angeles. Like New York City, Los Angeles has a strong mayor form of government: in addition to appointing members of the Board of Water and Power Commissioners, the mayor appoints additional managers and commissioners in the city (though they must be approved by the city council), can veto legislation, is responsible for policy implementation, and proposes the city's annual budget.

Angelenos often describe their city as a weak mayor city. This may have as much to do with the structure and powers of city council as it does with those of the mayor. In Los Angeles there are only fifteen council members, giving each councilor more authority and a broader constituency than their counterparts in other North American cities. Many Angelenos consider the city council to be the stronger body, with councilors presiding over their own individual "fiefdoms" within the city.[16] Between the mayor and the city council there has not been a unified effort to address climate change in the city until very recently,[17] and mayoral leadership carries less weight in Los Angeles than it does in New York City. Los Angeles has not historically been as involved as New York City with international networks like the C40, but this is changing under Mayor Eric Garcetti.

Los Angeles does not have significant financial capacities to leverage for climate change mitigation. According to city budget documents, the city has an operating budget of around $6 billion, which translates to $1,700 per capita. The city's relatively low revenues reflect a different set of responsibilities but are also the product of state legislation, Proposition 13, passed in 1978. This legislation prevents municipalities from assessing residential property at market value. Rather, property values are fixed to purchase price and increase with inflation, not more than 2 percent per year. As a result, property taxes make up a small portion of the Los Angeles budget, around 32 percent of general revenues. Other taxes and fees make up the remainder, including a transfer of 8 percent of LADWP revenue.

Los Angeles has a fraction of New York City's staff, with around 35,000 full-time city employees. Prior to the city's explicit commitment to reduce GHG emissions, staff had been engaged in internal energy efficiency and renewable energy programs. The city had constructed fifty new LEED-certified buildings, which they credit as helping "create a highly skilled workforce of architects, engineers, and contractors who have become experts in green buildings" (City of Los Angeles 2008). The city had also retrofitted buildings with energy efficient technologies and had begun to incorporate electric and alternative-fuel vehicles into their fleet. Very little integrated planning had been taking place, and in 2007 the LADWP had limited energy efficiency and alternative energy programs.

There are three pillars to Los Angeles city politics: business, labor, and environmental groups.[18] The Local 18 of the International Brotherhood of Electrical Workers represents the workers at the LADWP and are a potent political force in the city (Gottlieb et al. 2006). Their endorsement of mayoral candidates and support for new policies and programs hold significant sway. Environmental interests have played an important role in city politics since the election of Ruth Galanter to the city council in 1987 (Hughes, Pincetl, and Boone 2013). Los Angeles now prides itself on being an environmentally conscious city. The city developed one of the first green building codes in the U.S., and has some innovative water and wetland protection programs. Environmental interests are highly dispersed and decentralized; there are eighty groups in the city focused on the Los Angeles River alone.[19]

A unique feature of the City of Los Angeles is its position in a highly fragmented metropolitan area. The City of Los Angeles represents around one-third of Los Angeles County, which has eighty-eight individual municipalities, most of which have a population of fifty thousand people or more. This creates a need for cooperation on many services, including energy and transportation, and the County of Los Angeles plays a larger role in planning and regulation than in other places. The County has a $22.5 billion budget and has conducted multiple GHG inventories, developed an emissions reduction plan, and facilitates regional climate change initiatives such as the Los Angeles Regional Collaborative for Climate Action and Sustainability.

Los Angeles operates in the same federal context as New York City and has also taken advantage of small grants from federal agencies, but it has a much more engaged and proactive state government on climate change mitigation. In 2002 California developed a renewable portfolio standard that required publicly owned utilities to recognize the intent of the state legislature to encourage renewable resources. In 2006 the state emerged as a national (if not international) leader on climate change with the passage of AB 32, the California Global Warming Solutions Act. The law required the state to reduce GHG emissions to 1990 levels by the year 2020. In implementing AB 32, the state has introduced a series of additional legislation and guidelines. For example, in 2007 the state passed SB 1, which requires that municipal utilities (like the LADWP) develop a solar rebate program. In 2008 the state passed SB 375, which requires that regional planning organizations such as LA Metro develop plans for meeting regional GHG emission reduction targets. California introduced a cap and trade system for carbon emissions in 2012 that is now linked to Quebec. In 2018, the state passed unprecedented legislation to commit to powering the state entirely through renewable energy resources and zero-carbon resources by 2045.

While California clearly provides valuable leadership and resources that Los Angeles can use to pursue its climate change goals, in 2007, like *PlaNYC*, the *GreenLA* plan was a city-led initiative. The city was not expected to produce a climate change plan, and there is very little in the plan that was already required of the city by the state. This relationship has shifted over time as the state's programs have become more ambitious and regulative, and the city has struggled at times to build and maintain support for its GHG emission reduction strategies. Air quality, traffic congestion, and long commuting times are perennial issues for Angelenos and their representatives.

Toronto

Similar to New York City, Toronto conducts a GHG emission inventory on a fairly regular basis. In 2007 the city reported its total emissions for 2004 as 23.4 million metric tons, or around 9 tons per person (City of Toronto 2007b). It attributed around one-third of these to transportation, 12 percent to waste management, and 55 percent to energy used in buildings. The city's emission profile is similar to Los Angeles and reflects a similar urban morphology. Toronto is Canada's largest city, with 2.8 million people in the city's boundaries. With a population density of around 10,000 people per square mile, Toronto is about 20 percent denser than Los Angeles but still has half as many people per square mile as New York City. This moderate density, combined with chronic underinvestment in transit, has led Toronto to become a fairly car-centric city: 70 percent of the city's residents drive to work each day (Statistics Canada 2011). The city's most recent emission inventory (released in 2015 using 2013 data) shows an increase in the contribution of transportation to 41 percent, but the city has struggled to update its transportation figures since 2008 (City of Toronto 2015). Toronto plans to begin reporting using the GCP for the next inventory, which may also shift the city's estimates slightly.[20]

Also similar to New York City, Toronto's energy system and infrastructure is largely managed and delivered by a regional network of public and private entities. Ontario deregulated the electricity sector in 1999. The restructuring produced an independent Crown corporation, the Independent Electricity System Operator (IESO), which manages the province's electricity supplies as well as the Ontario Power Authority, a long-term planning body that develops and delivers energy conservation programs throughout the province (later absorbed by the IESO). Ontario Power Generation produces much of the province's electricity. Natural gas for the city comes from a private company, Enbridge. Toronto Hydro, a municipal corporation owned by the City of Toronto, owns and operates the city's electricity distribution system and collects fees and is governed by a board of directors composed of eight city council appointees, the mayor, and two city councilors.

In 2002, Ontario Premier Ernie Eves pledged to shut down all coal-fired power plants, and in 2014 Ontario became the first jurisdiction in North America to be coal-free. This transition was largely driven by concern for smog in the region and the general deindustrialization of Ontario's economy. As an officer from the Ministry of Energy put it, "the coal plants were old and inefficient to begin with; it made for an ambitious signature project."[21] In 2007, around 40 percent of Toronto's electricity supplies were produced from coal, the city had close to zero renewables, and natural gas for heating accounted for 60 percent of the city's energy use–related GHG emissions.

The Toronto city government and mayor are the weakest of the three cities. In Canada, cities are considered to be creatures of the province, and the Ontario government plays a large regulatory and planning role for the city. Canadian cities lack many of the policymaking and taxing powers that U.S. cities enjoy, including the ability to set local building codes and standards. These powers have gradually been expanding for Toronto as the city has grown and sought to gain more control over its destiny. In 2006, under pressure from Toronto, the province passed the City of Toronto Act. The act expanded the powers of the city to include some additional taxing powers (transit tax, hotel tax) and the ability to pass bylaws that protect local health or environment. The city has yet to use many of these new powers, and it can often be difficult to tie new initiatives directly to local health or environment.[22]

A unique feature of Toronto is the relative authority it has over its transportation system. While funding for large projects often comes from (at least in part) the provincial and federal governments, city council has the authority to plan for and authorize new transportation projects like light rail, subway lines, and rapid bus service. In recent years this has tended to produce a growing list of unfunded but council-approved transit projects. Regional transit networks are governed through Metrolinx, a provincial transit agency. The city's transit network is operated by the Toronto Transit Commission (or TTC), a public city agency governed by an appointed CEO and board of commissioners. Toronto is also unique in the relatively high proportion of GHG emissions it attributes to waste management (between 10 percent and 12 percent), and here (like most cities) the city also has significant authority over waste hauling and processing practices and facilities.

Initial political leadership on climate change came from former Toronto mayor David Miller, who was in office from 2003 to 2010 and now serves as the North American director for the C40 Cities Climate Leadership Group. He led the development of the city's climate change plan and invested political capital into moving the initiatives through city council. Since Miller left office in 2010, the city has not had a clear political champion for climate change. There have been consistently supportive city councilors, but their efforts to forward a climate change agenda legislatively have been sporadic and often solitary. Toronto city government is a distinctly weak mayor system, so political leadership in city council is crucial. The mayor has very little formal power to control the city's policy agenda; he or she has one vote in the city council, so the ability to advance city-wide initiatives is dependent on more informal uses of authority and persuasion.

The political possibilities for climate change policy in Toronto are heavily influenced by the opportunities present in city council.[23] Amalgamation has changed the city's politics significantly by combining relatively conservative, car-dependent post-war suburbs with a liberal, wealthy financial district downtown (Keil 2000; Horak 2013). In 1998 the City of Toronto went from one of six municipalities under the Metropolitan Toronto government to an amalgamated megacity. The two-tiered system was transformed by the provincial government to a single-tiered municipality with fifty-six (later forty-four) city councilors, despite significant opposition from residents (Horak 2013). Today the City of Toronto is home to 2.8 million people, which is one-fifth of Ontario's total population.

Toronto's post-amalgamation city council is large (forty-four members), nonpartisan, and the commitment since 2007 of both mayors and city council to climate change mitigation has been highly variable. Many of the city's political battles still fall along pre-amalgamation downtown–suburban divides, and climate change mitigation is viewed by many as a distinctly downtown agenda. While Mayor David Miller was able to bridge some of these divides during his time as mayor and gain support for climate change mitigation, such coalition building was short lived. The absence of political leadership on climate change in Toronto has diminished the city's role internationally. For example, while David Miller served as chair of the C40 from 2008–2010, Mayor Rob Ford (2010–2014) did not attend meetings and John Tory has been to just one.

The political dynamics of climate change policy in Toronto stand to shift significantly following the election of a Conservative provincial government in 2018 that positioned Doug Ford as Ontario's premier. Ford has initiated a plan to shrink the Toronto city council from forty-seven to twenty-five, and there is evidence that his party has banned provincial agencies from referring to climate change in their communications (O'Neil 2018).

Toronto's financial capacities fall between those of Los Angeles and New York City. According to city budget documents, the city has a $10 billion operating budget, or around $3,700 per capita. In recent years Toronto has faced serious budget shortfalls, largely due to a longstanding political commitment to keep property tax increases at the rate of inflation, and is currently grappling with options for increasing revenues given institutional and political constraints.[24]

The primary strength Toronto brings to climate change mitigation is its administrative and technical capacities. Toronto was an early actor on climate change. When it first pledged to tackle GHG emissions in 1990 it also created two important institutions: the Energy Efficiency Office (now the Energy and Environment Division) and the Toronto Atmospheric Fund (now The Atmospheric Fund). Both institutions launched programs in the 1990s that continue to shape Toronto's mitigation efforts today. The Energy Efficiency Office created the Better Buildings Partnership in 1991, one of the first municipal efforts to reduce energy use in large buildings. It also helped to launch a network of deep lake water cooling systems downtown, which significantly reduces energy used to cool large buildings and is an important innovation in local, sustainable energy systems. These initiatives in the Energy Efficiency Office have helped to generate experience and capacity with emission-reducing activities that the other two cities didn't have.

Perhaps most important for the city was the creation of the Toronto Atmospheric Fund (TAF), a type of in-house think tank and innovation catalyst that has provided an opportunity for the city to experiment and innovate. In 1991, led by a progressive coalition of councilors, the city sold some its assets and half of the proceeds ($25 million) went to an endowment for TAF and its role in supporting GHG emissions and air pollution reduction actions for the city of Toronto. TAF is therefore financially independent from the city, not funded by tax revenues, and does not deliver programs for the city. Rather, its role is to do "impact investment"; its role is to "develop and de-risk programs for the city that can be scaled up."[25] It provides financing to energy and GHG-related projects that would otherwise have trouble getting off the ground. These have included an initial grant to the city for LED lights, developing and delivering new energy efficiency financing tools, and helping to lead public and stakeholder engagement activities. TAF often tries to target pilot projects and high-impact investments, serving as a bridge between the city and climate change champions. Unlike city departments, it understands it has latitude for programs to fail and is willing to take on more experimental initiatives and help incubate new ideas.[26] This has served as an important source of capacity for Toronto as it has worked to reduce GHG emissions.

Similar to Los Angeles, Toronto has benefitted until recently from a relatively proactive provincial government.[27] After phasing out coal, the province pledged to reduce emissions 15 percent by 2020, 37 percent by 2030, and 80 percent by 2050. As part of its plans to reach these targets, the province launched in 2017 a cap and trade program for GHG emissions. The first cap and trade program auction generated nearly C$500 million for the province, and these funds must be reinvested in programs that will further reduce GHG emissions in the province (Ontario Ministry of Environment, Conservation, and Parks 2017). The province has also consistently provided funding for energy conservation and research and development for energy efficiency technologies. However, in July 2018 the Ford administration canceled the provincial cap and trade program.

Canada's federal government has historically failed to develop significant climate change policy, withdrawing from the Kyoto Protocol in 2011. Under Prime Minister Trudeau, federal climate policies seem to be taking a turn. The prime minister committed to the goals of the Paris Climate Agreement and announced in 2016 a forthcoming pan-Canadian cap and trade program. Like the province's recent efforts, it is still to be seen what effects these shifts will have for Toronto's own work to reduce GHG emissions.

New York City's Policy Agenda for Reducing GHG Emissions

New York City's sweeping sustainability and climate change plan, *PlaNYC 2030: A Greener Greater New York (PlaNYC)*, has been heralded as "a breakthrough in sustainability planning, and is considered by local governments around the country to be the gold standard for big-city sustainability plans" (ICLEI 2010). By 2007 the city had successfully rebounded from the economic and social troubles it faced in the 1970s, diversifying its economy, reducing crime, and adding nearly 1 million new residents (Glaeser 2011; Solecki 2012). With these changes had come steady population growth, increased energy consumption, a commuter shed that spanned more than 10,000 square miles, and 55 million tons of GHG emissions each year. In response to these stressors, Mayor Michael Bloomberg initiated a strategic planning process that ultimately centered on long-term sustainability and climate change. The city spent 2006 holding public consultations and meetings with stakeholders and assembling the administrative apparatus necessary to develop and implement a large-scale sustainability plan.[28] On Earth Day, 2007, New York City released *PlaNYC 2030*, aiming to reduce the city's GHG emissions 30 percent below 2005 levels by 2030. In doing so, the city also sought to serve as a model for what the economically competitive, environmentally sustainable city of the future could be.

Bloomberg's commitment to climate change was a major turning point for the city. Many city council members had long been interested in climate change mitigation but their efforts had lacked support from the mayor's office (Climate Group 2005; ICLEI 2010). According to Marion Ariella, a senior policy advisor at the time, "nobody was guiding the ship" (quoted in Bagley and Gallucci 2013). This changed with Bloomberg's commitment to climate change and the election of Christine Quinn as city council speaker in 2008. Bloomberg and Quinn became fast political allies and collaborated heavily to develop the programs and policies necessary to reduce GHG emissions. In fact, "no one had collaborated like this with the administration on an agenda before."[29]

Bloomberg's influence also extended beyond structural powers, as he held significant sway with the city's business community. He had made his fortune in the private sector and is a highly respected businessperson in the city, giving him credibility with the business sector that other mayors have not always enjoyed. He leveraged this influence to promote his climate change agenda: "If you wanted Bloomberg to speak at your event, do something with sustainability and he'll announce it."[30]

When Bill de Blasio took office in 2014, it was not immediately obvious that he was committed to continuing the city's work on climate change. He positioned himself during the mayoral campaign as the "anti-Bloomberg" candidate and likely saw little value initially in championing one of Bloomberg's signature initiatives. Under pressure from environmental groups and the international community,[31] the de Blasio administration released in September 2014 its own ambitious plan for climate change mitigation, *One City, Built to Last: Transforming New*

York City's Buildings for a Low-Carbon Future. De Blasio's plan introduced a new goal of reducing GHG emissions 80 percent below 2005 levels by 2050.

One City builds on the successes of *PlaNYC*, aiming to expand many of the Bloomberg administration's key initiatives. Importantly, it also expands the agenda by linking climate change mitigation to de Blasio's commitment to income inequality and housing affordability. The administration's subsequent overarching plan for the city, *One New York: The Plan for a Strong and Just City*, presents both a vision for climate change mitigation in New York City and ambitious social policies such as raising the minimum wage and lifting 800,000 New Yorkers out of poverty by 2025.

New York City's subsequent policy agenda for reducing GHG emissions has made good use of the city's capacities and sources of authority but, more centrally, it has leveraged the city's strong political leadership and commitment to climate change.

Regulating Energy in Buildings

The Bloomberg administration, working closely with the city council, focused on regulating energy in buildings in three ways: changing municipal building codes, requiring energy-use reporting, and banning highly polluting heating oils. Through these efforts the city has passed more than one hundred pieces of legislation since 2007.[32] Bloomberg's leadership on regulating energy in buildings lent the work credibility and helped foster buy-in. The shared commitment in the city council allowed them to work in lock step with the mayor. "On the building side we are fortunate because we have authority so we try to do what we can do. We have the political will and a strong mayor . . . we try to be opportunistic but we really focus where we have control."[33]

First, the city used municipal building codes to improve energy efficiency in buildings. Targeting the codes, they thought, would lower compliance costs by creating economies of scale for expertise and materials, is enforceable, and builds on existing institutions and industry practices. One important piece of legislation to come out of this effort was Local Law 85 of 2009. This established a city-level energy conservation code that superseded those of the state. The New York City Energy Conservation Code closed a loophole stipulating that renovations that affect less than 50 percent of a building didn't have to comply with the code; now the renovation of any portion of a building must conform to building code provisions for new construction.

All told, the city council has adopted at least fifty-three updates to various municipal codes to improve energy efficiency. These include very broad changes, such as introducing environmental protection as a fundamental principle of the construction codes, and very specific changes, such as insulating exposed pipes during construction. The measures go beyond LEED certification standards for energy efficiency measures and incorporate social equity goals. The C40 calls it "the most comprehensive effort of any U.S. city government to green local construction codes and regulations that impacts buildings" (C40 2012a).

Second, New York City developed a set of standards and requirements for the operation of buildings through the Greener, Greater Buildings Plan (GGBP). This program is considered by many to be the city's greatest achievement, and various parts of it have been replicated by cities around the world. In December 2009 Bloomberg signed a package of legislation called the Greener, Greater New York Buildings package that codified various aspects of the GGBP. Two of these measures, Local Laws 85 and 88, set new requirements for building owners, including bringing large buildings up to the city's energy code by 2025.

Finally, the city passed legislation banning the use of heavy polluting heating oils in buildings (Numbers 4 and 6). Local Law 43 put a 1,500 ppm cap on the amount of sulphur that comes from heating oils and required that all heating oils in New York City contain 2 percent biodiesel. This requires building owners switch to cleaner-burning heating oils, which has often required replacing boilers or furnaces in the process.

New York City has complemented its regulatory approach to managing energy use in buildings with the voluntary Mayor's Carbon Challenge. Bloomberg challenged the city's universities and hospitals to match the city government's accelerated target of a 30 percent reduction in GHG emissions by 2017. "Just as the city was poised to lead, we thought there were certain sectors poised to lead."[34] Mayor de Blasio expanded this program to include multifamily residential buildings in August 2014, and the city is now working with eighteen of the largest residential property management companies toward accelerated GHG emission reductions (Tatum 2014).

An Adaptive Transportation Agenda

One of the first initiatives to come from *PlaNYC* was an effort to introduce congestion pricing in the commercial center of Manhattan—a regulatory approach to targeting GHG emissions from cars. Congestion pricing charges motorists a fee to enter the downtown business district and is "designed to encourage shifting peak period trips to off-peak periods, away from congested facilities, and onto alternative transportation modes" (Dahlberg, O'Brien, and Zoller 2007, 2). Congestion pricing was implemented in London in 2003, but this was the first time it was seriously considered in a North American city (Schwartz et al. 2009). The city wanted to pilot congestion pricing for three years and had a financial commitment from the federal government of $354 million to support the effort. The pilot would apply between 6:00 a.m. and 6:00 p.m., and drivers would pay an $8 daily fee to enter the zone (trucks would pay $21). The city anticipated the fee would result in a 6.3 percent reduction in traffic and a 7.2 percent increase in speed, helping to reduce transportation-related GHG emissions.

The challenge for New York City was that introducing a congestion charge required legislative approval from the state government. Bloomberg lobbied heavily for state support and gained backing from the city council and business, environmental, civic, and transportation advocacy groups. The state formed a special commission to study the proposal, and it was eventually introduced into the state legislature for approval. Several legislators expressed serious concerns, which primarily hinged on the potential for the program to target low-income commuters and the city's inability to guarantee—because of its limited authority over transit—that revenues from the program would go toward improved public transportation. The state legislature ultimately prevented the proposal from reaching the floor and the initiative failed. In 2017 Governor Cuomo revived the congestion pricing debate but it was once again unsuccessful in the state legislature and opposed by Mayor de Blasio, "despite months of lobbying by advocates, a six-figure media campaign, and rallies by transit riders" (Hu 2018). In March 2019 the state legislature passed a measure allowing a $10 fee to be charged for driving in the busiest parts of the city.

The city experienced a similar defeat with its attempt to regulate emissions from taxis. In 2007 the city's Taxi and Limousine Commission developed rules requiring that all taxis in the city be hybrid vehicles by 2012. The initiative met with significant resistance from the Metropolitan Taxicab Board of Trade, the second most powerful lobbying group in the city.[35] In 2008 the requirements were challenged in state court, where it was determined that it was outside of NYC's authority to establish mileage standards for private vehicles.

The obstacles encountered by the city in its work on reshaping the transportation system were not unexpected given the city's limited authority—"the transportation piece was a Hail Mary."[36] When a regulatory approach to transportation emissions failed, the city adopted a more provisioning-oriented strategy, largely under the direction of Transit Commissioner Jeanette Kahn. Rather than a single flagship program, the city is pursuing multiple reorientations of the city's infrastructure to facilitate alternative modes of travel. For example, the city has required that 20 percent of new parking be charger ready for electric vehicles and has added over seven hundred new miles of bike lanes (City of New York 2013). The New York City DOT developed a "Sustainable Streets" strategy that promotes multiple modes of transportation, increasing public space, road safety, and bus rapid transit. Both Bloomberg and de Blasio have closed different parts of the city to cars, including Times Square and parts of Central Park. Learning from its failed attempts to directly regulate emissions from cars, the city is seeking to reorient city services and infrastructure in ways that facilitate a less emissions-intensive transportation system.

Enabling Changes to the City's Energy Supplies

Changing energy sources is an effective way of significantly reducing urban GHG emissions, particularly in New York City where energy use accounts for nearly 80 percent of the city's total emissions. It is difficult for the city to target this emissions source because they have very little formal control over energy supplies; bringing energy supplies into the city isn't its job.[37] What the city has done instead is taken steps to make it easy and attractive for other actors in the energy system, such as ConEdison, to move away from coal and incorporate lower emission energy sources, particularly natural gas and solar power.

Prior to *PlaNYC* the city relied on natural gas for around half of its energy use, and the city sought to expand the use of natural gas even further to reduce emissions. Given the limited regulatory tools available for this, the city focused instead on supporting and incentivizing the increased use and accessibility of natural gas in the city and the improved efficiency of electricity generation in the city's power plants. The city collaborated with ConEdison, using memorandums of understanding, to make it easier for New Yorkers to access natural gas by expanding the natural gas distribution system in the city and building a new natural gas plant in Astoria.[38] The city went as far as to offer to pay ConEdison to expand natural gas distribution systems; but so far hasn't needed to, as demand has been relatively high.

The Public Service Commission's annual "rate cases" are important venues for interaction and negotiation between ConEdison and New York City.[39] At these rate case hearings, the city can formally advocate for measures it would like to see taken to reduce energy demand or increase renewable energy supplies. The city has used these venues and others to advocate for and support the repowering of two of ConEdison's steam plants that now run exclusively on natural gas.

New York City has also focused on expanding renewable energy supplies for the city. In 2008 the city passed legislation allowing property owners to deduct up to 20 percent of the cost of solar power systems from their property taxes and updated the building code to allow the land use changes necessary for solar power installation on residential and commercial buildings. Mayor de Blasio has taken up the issue more aggressively, initiating a number of programs aimed at helping to bring more solar energy to the city. In 2014 he pledged to install 100 MW of solar power on city-owned buildings by 2025, and committed $23 million to installing solar panels on city-owned buildings.

The larger impact may ultimately come from the city's ability to choose its own source of energy for government operations given the deregulated, largely private energy market in New York. As a very large energy purchaser (responsible for 80 percent of regional and 50 percent of state energy demand), the city can indirectly influence the energy market by supporting alternative energy sources.[40] In July 2015 de Blasio pledged to use the city's purchasing power for electricity to ensure that 100 percent of city operations will be powered with renewable energy as part of the city's goal to reduce GHG emissions 80 percent by 2050. That year the city released a Request for Information document regarding possibilities for new generation capacity able to power city government operations entirely with renewable energy sources. The aim was to motivate private enterprises to think more seriously about how to get renewable energy to New York City (Giambusso 2016). De Blasio said, "This is a call to the marketplace: the biggest energy customer you'll find is ready to put our money where our mouth is when it comes to renewable power" (Office of the Mayor 2015a).

The proposals the city received seem to have laid bare some of the challenges to meeting the goal despite the deregulated market: long-term contracts for natural gas, the need for project approval and development from other jurisdictions, and the city's ongoing relationship with the New York Power Authority. The de Blasio administration sees a way forward using a mix of wind, solar, and hydropower and working in cooperation with the state, which has now also adopted a goal of reducing GHG emissions 80 percent by 2050 (Giambusso 2016). Powering New York City's municipal operations using entirely renewable energy sources would certainly have an impact on renewable energy capacities and technology in the state and possibly the region.

Los Angeles's Policy Agenda for Reducing GHG Emissions

Los Angeles's commitment to climate change is an outgrowth of the city's long-standing political commitment to environmental issues and increasingly sophisticated and politically influential environmental interest groups. Together with business and labor, environmental groups are considered to be "one of the three legs of political power in City Hall."[41] It was the emergence of a broader coalition of environmental, environmental justice, social justice, and health advocates in the 1990s—the Progressive Los Angeles Network (PLAN)—that helped pave the way for climate change to rise on the city's agenda.

Leading up to the 2001 mayoral election, the PLAN forwarded a twenty-one-point agenda for social and environmental progress that included public transportation, green energy, and clean fuel vehicles. Mayoral candidate Antonio Villaraigosa, then a California assembly member with deep ties to both the environmental and labor communities, took up the agenda and actively engaged with the environmental community during the election. He ultimately lost to James Hahn, but in 2005 ran again and won.[42] Villaraigosa's victory signaled a victory for the PLAN's progressive agenda and for labor and environmental groups in particular. LADWP union leader Brian D'Arcy wrote in a 2005 union newsletter: "on May 17 we helped elect *one of our own.*"

Villaraigosa's victory helped to further coordinate and consolidate the diverse set of actors and organizations that had supported his mayoral bid and provided environmental groups with greater access to the city's policymaking process.[43] Environmental supporters—now located both inside and outside of city government—subsequently formed the Green LA coalition, a broad-based coalition of over one hundred environmental and environmental justice groups in the city.[44] Together, the Green LA coalition and the mayor developed an environmental plan for the city that was explicitly focused on addressing climate change in Los Angeles and called *Green LA: An Action Plan to Lead the Nation in Fighting Global Warming,* committing to reduce the city's GHG emissions 35 percent below 1990 levels by 2030, in the process making Los Angeles "the greenest city in America" (City of Los Angeles 2007). This is a slightly more ambitious target than New York City's goal of a 30 percent reduction by 2030, but uses 1990 levels as its baseline, which may not have been the city's peak levels.

Villaraigosa was replaced by Mayor Eric Garcetti in 2013. As city council president, Garcetti had been critical of the Villaraigosa administration, and he ran on a "back to basics" agenda focused on improving services and encouraging job growth while the city began to recover from the recession. Leading up to the election, Mark Gold of UCLA's Institute of Environment and Sustainability drafted a vision of a sustainability plan for the city called *Vision 2021: A Model Environmental Sustainability Agenda for Los Angeles' Next Mayor and City Council.* The vision aimed to provide Los Angeles the kind of comprehensive sustainability plan pioneered by New York City, and was taken up by both Garcetti and his opponent, Wendy Greul, during the campaign.

After being elected, Garcetti developed a sustainability plan that met all of the criteria put forth in Gold's *Vision 2021.* On April 8, 2015, the mayor released the *pLAn,* a wide-ranging sustainability plan for Los Angeles that presents climate change targets alongside strategies for improving the city's economy and addressing equity issues. The *pLAn* commits the city to reducing GHG emissions 45 percent below 1990 levels by 2025, 60 percent by 2035, and 80 percent by 2050.

Outside of the formal planning process, the city has repeatedly committed to phasing out coal entirely. In his second inaugural address in 2009, Villaraigosa claimed the city would be off coal by 2020 and getting 40 percent of its energy from renewable sources. In 2013 Garcetti led a motion in the city council to formalize this commitment, and in 2016 the council directed the LADWP to determine how to move the city to 100 percent renewables. Councilors have framed this recent initiative as a way to reduce air pollution, avoid the disastrous consequences of climate change for coastal areas, and foster the clean energy economy in Los Angeles (Kinman 2016).

Los Angeles's policy agenda for GHG emission reductions has largely leveraged its valuable authority over energy supplies and demand via the LADWP. Combined with a focus on new transportation infrastructure, this has resulted in a highly provisioning mode of governing GHG emissions in the city.

Providing New Energy Services

As the city works to meet its climate change mitigation goals, Los Angeles consistently draws on its ability to directly shape energy services, develop renewable energy, and promote energy conservation.

A central focus for Mayor Villaraigosa was fulfilling at least 20 percent of the city's energy needs through renewable sources by 2017. In the initial stages of developing renewable energy, the city focused heavily on wind: "at the time, wind was far more economical than solar power and the industry was more mature" (Villaraigosa, Sivaram, and Nichols 2013). The city has been successful in bringing wind power on board, and by 2010 it accounted for nearly 20 percent of the city's electricity consumption. As the price of solar energy began to drop and state and federal incentives for solar came on board, the city has switched its emphasis to developing solar power.[45]

Solar power is an intuitive choice for a city that receives more than three hundred days of sunshine per year. As a first step, the Villaraigosa administration released a plan in 2008 called *Solar LA*, which lays out a strategy for the city to develop 1,300 MW of solar energy capacity. There are three components: (1) customer solar programs (380 MW), for which the LADWP uses incentives to encourage residents and private developers to install solar panels on residential and commercial buildings; (2) a city-owned solar network (400 MW), for which the LADWP would own and install solar power systems on city-owned property; and (3) large-scale solar projects (500 MW), for which the city would purchase solar energy from large-scale projects developed outside of the city, most likely in the Mojave Desert. While the large-scale solar projects are generally more economical, the *Solar LA* plan prioritized the job creation potential of local solar projects. The city estimated that every 10 MW of locally generated solar energy would create two hundred to four hundred jobs in the city. Developing local solar power helps to demonstrate that "we're doing it here too, not just buying land elsewhere and then building transmission lines through."[46] Environmental groups have also criticized large-scale solar projects as being an "industrialization of the desert."[47]

Los Angeles has used two common tools for promoting solar energy: a Solar Incentive Program and a feed-in tariff model. The Solar Incentive Program provides rebates to home and business owners who install solar rooftop programs. The LADWP developed a ten-step declining rebate based on the capacity of the system installed and connected to the grid, starting at $3.25 per watt in 2007 and declining to $0.30 per watt in 2016. This was one of the highest incentive levels in California (above market price) and coincided with a 30 percent rebate for solar energy installation from the federal government through the residential renewable energy tax credit.[48] One outcome from this program is the rapid growth of the solar energy installation and operation company Solar City, which owns 70 percent of the rooftop solar installations in Los Angeles.[49]

The LADWP also introduced a feed-in tariff program in 2013, which provides a mechanism for third parties to generate solar power and sell it to the Los Angeles Department of Water and Power. This is not a residential program but rather is intended to encourage solar developers to build medium-sized solar energy systems in the city. This was developed in part as a response to state legislation (SB 1332), which required utilities to provide a feed-in tariff system and specifically 75 MW of capacity from the LADWP. The LADWP went beyond state legislation and set a goal of 150 MW for the feed-in tariff system and, again in order to increase participation, set the incentive levels above wholesale energy rates. The city rolled out the program incrementally, with five semiannual allocations of various sizes. The first 100 MW came with a guaranteed rate of $0.17/kWH (accounting for the avoided cost of large-scale solar development and transmission, and economic benefits of local solar) to allow developers to evaluate project options and demonstrate a long-term commitment from the LADWP.[50]

Prior to 2007, the LADWP was not investing significantly in energy efficiency and conservation. As both the city and the state began committing to reduce GHG emissions, efforts to fund and promote energy efficiency in Los Angeles also ramped up. By 2012 the utility had a $267 million budget over two years for energy efficiency programs, at the time an all-time high for the LADWP. It later developed a strategic plan for investing between $100 million and $200 million annually in energy efficiency and conservation from 2015 to 2020. The programs through which these funds are funneled range from incentives for small businesses, to outreach and partnership efforts, to refrigerator recycling opportunities.

In 2014 the state mandated a 10 percent reduction in energy use by 2020 for energy utilities, including the LADWP. City council members and the Board of Water and Power commissioners took the step of explicitly going beyond this target and adopted a goal for the city of reducing energy use 15 percent by 2020, claiming it was "the most ambitious energy efficiency goal by a major municipal utility in the United States" (Los Angeles Office of the Mayor 2014). The city sees significant advantages to adopting the more aggressive target, and generally to adopting a conservation-oriented approach to energy services.

David Jacot, director of energy efficiency programs at the LADWP, finds at least four reasons for the city to prioritize conservation: "to demonstrate leadership consistent with being the largest and arguably most innovative city-owned utility in the U.S.; as a least-cost compliance strategy for meeting state GHG emissions reduction targets in the mid- and long-term; as a least-cost supply source for replacing coal and meeting renewable portfolio standards; and as the most potent local jobs creator of any supply-side resource."[51] This view is increasingly shared by a range of stakeholders and decision makers in the city. According to Jacot, "politically, efficiency is easy."

Providing New Transportation Infrastructure

For New York City, transportation represents just 20 percent of the city's GHG emissions; but in Los Angeles this figure is at least 34 percent, and in 2007 the city believed transportation to be responsible for nearly half of city-wide GHG emissions. Villaraigosa subsequently prioritized transportation during his time as mayor and initiated a number of projects within the city designed to provide transportation services and infrastructure compatible with the city's goals for reducing GHG emissions. For example, the city synchronized stoplights to improve traffic flow and the LADWP has installed electric vehicle charging stations throughout the city.

Restructuring the city's transportation system requires larger-scale investments and transformations, which in turn requires significant capital and regional coordination. To achieve these, Los Angeles leveraged its influence within Los Angeles County and the LA Metro board to secure significant new revenue for transportation projects that would help to reduce the city's GHG emissions, improve air quality, and ease congestion.

In 2008 Villaraigosa played a leadership role in championing Measure R, a county-level ballot measure that introduced a 0.5 percent increase on local sales taxes in order to pay for transportation projects in the region. Projects include extending subway lines to new parts of the city and county, including the "Subway to the Sea" that runs from Culver City to Santa Monica; a rail extension to the Los Angeles International Airport; and a light rail system to connect subway lines via downtown Los Angeles. Villaraigosa was a very active campaigner for the ballot measure, investing significant capital into this effort, in part because it would help the city of Los Angeles to meet its GHG reduction targets. The measure required a two-thirds supermajority and ultimately passed with a narrow 0.4 percent margin.

The sales tax increase is set to generate $40 billion over thirty years. Nearly 80 percent of this revenue is earmarked for public transit capital and operations investments. Indeed, the city explicitly sees itself as "moving away from freeway-based solutions" as it charts out a new path for transportation infrastructure.[52] The list of projects proposed to be funded by Measure R revenue is long, in part to increase political support, and likely cannot all be paid for using the sales tax increase alone.

It quickly became clear that while the revenue from Measure R needed to accumulate over time, the city's transportation problems and GHG emissions were not slowing down. Villaraigosa therefore went a step further, successfully lobbying the federal government for rule changes for transportation funding that allow projects supported by Measure R to be built within ten years. In 2011 the U.S. Congress passed the *Moving America Faster* bill, which provides loans for transportation projects that can be paid off as revenue accumulates over the thirty-year lifespan of the sales tax increase. The bill ultimately passed with the support of the U.S. Chamber of Commerce, national union representatives, and over sixty mayors (Newton 2011). This has helped significantly to jump-start transportation investments in Los Angeles.

Still, the city and county felt financially constrained in the face of the significant investments in transportation needed in the region. In order to further accelerate investments in transit infrastructure, the county has twice attempted to replicate its success with Measure R. In 2012 voters rejected a measure to extend the Measure R sales tax increase by an additional thirty years. Again the decision hung on a narrow margin, with Measure J receiving just 66.4 percent approval. In 2016, voters approved an additional and permanent $0.05 sales tax increase to fund transportation projects (Measure M). This time voters were more decisive, with around 70 percent voting yes. Like Measure R, the measure includes an independent taxpayer oversight committee and commits just 17 percent of new revenue to highway construction projects. Taken together, these new revenue sources have the potential to transform transportation services in Los Angeles. Having the opportunity to pursue these options locally through the county and LA Metro meant that Los Angeles could leverage its influence in both arenas to forward progressive transportation goals.

Toronto's Policy Agenda for Reducing GHG Emissions

Toronto was a climate change pioneer, becoming in 1990 the first city to make a formal pledge to reduce its GHG emissions. At the time the city aimed to reduce emissions 20 percent below 1988 levels by 2005 (Harvey 1993). Concerted efforts to pursue these goals stumbled as the city underwent a tumultuous period of amalgamation and reconfiguration.

Elected in 2003, David Miller was the city's first progressive mayor postamalgamation and the city's first mayor to take up climate change. Miller was endorsed by then-Toronto resident Jane Jacobs and had built a coalition of downtown councilors and some progressive suburban councilors based on shared interests in transit and housing. He spent his first term trying to get control of the new megacity, and environmental issues were somewhat sidelined.[53] When he was reelected in 2006, environmental advocates like the Toronto Environmental Alliance called on Miller to take climate change more seriously, and it became a priority for his campaign and administration.

The 2006 election also brought a wave of progressive city councilors, including Gord Perks, who had spent his career at the Toronto Environmental Alliance.[54] Subsequently, for both Miller and the city council, climate change was one of the top three items for that term. In 2007 the city released *Change is in the Air*, a plan to reduce GHG emissions in Toronto 6 percent below 1990 levels by 2012 (what they called the "Kyoto target"), 30 percent by 2020, and 80 percent by 2050 (City of Toronto 2007a). This was followed in 2009 by a detailed sustainable energy strategy called *The Power to Live Green*.

In 2010 Toronto elected populist mayor Rob Ford, who represented a backlash from the city's inner suburbs against a perceived progressive, downtown-driven agenda. Ford didn't prioritize or target the city's climate change work but was known to vilify people and programs he didn't like in the press, referring to environmental groups as special interests.[55] The size and unlikelihood of his victory made moderate councilors closed to environmental issues, as they didn't want to invite an attack from Ford.[56] City staff in the Environment and Energy Division were able to "fly under the radar" during Ford's tenure and implement existing programs without building anything new.[57]

With the decline of Ford's influence at the end of his term, and the election of John Tory in 2014, the city has been regaining some momentum on climate change. This shift had more to do with ongoing support from the city council than mayoral leadership. While Tory publicly recognizes the importance of climate change—attending the Paris Climate Conference in 2015—he hasn't made it a priority of his administration. Supporters in the city council felt empowered to revisit the issue and renewed their commitment to climate change mitigation. While work had continued during the Ford administration, strategic planning and monitoring had fallen to the wayside. The council subsequently created a subcommittee on Parks and Environment for climate change mitigation and adaptation, and in March 2015 tasked the Energy and Environment Division with determining whether the city's existing plans will achieve long-term GHG emission reduction targets. With the answer likely being "no," the city is subsequently exploring policies and programs able to reduce the city's GHG emissions 80 percent by 2050. The effort is called *TransformTO* and is unique among city climate change plans in that it aims to use multicriteria analysis scenario development to explore alternative pathways to achieving 80 percent emission reduction and allow stakeholders to develop and explore these alternative scenarios. The plan also links the city's work to reduce GHG emissions with broader goals to improve livability, health, and equity in the city and includes an extensive public engagement agenda. While *TransformTO* has so far received incremental funding from the city council, it is making progress.

The policy agenda Toronto has developed since 2007 for reducing GHG reductions has relied primarily on its internal capacities to develop innovative approaches internally and enable action by other actors. The city's work has largely been driven by the commitment of the public service, including The Atmospheric Fund (TAF).

Enabling Energy Retrofits

Since 2007 Toronto has developed a number of tools for enabling action by commercial, institutional, and residential building owners to reduce energy use. This has often involved creatively leveraging the limited formal powers of the city to access new sources of revenue and opportunities for incentives.

The city wanted to show leadership on climate change and saw an opportunity for it to act as a source of financing for building energy retrofits. In 2007 the city partially monetized outstanding debt from Toronto Hydro to create a $62 million fund to support building retrofits—the Sustainable Energy Funds—and establish the city's Energy Efficiency Office.[58] The funds have been used as a revolving loan program for retrofits in both city-owned and privately owned buildings. The idea came from the finance department, which was eager to contribute to the city's efforts to reduce GHG emissions.[59] The money was loaned interest-free as a way of assisting city agencies and nonprofit groups that had trouble accessing capital for such projects.[60]

By 2012, $28.6 million in loans had been made, primarily to city government entities (agencies, divisions, corporations of the city, business improvement areas), hospitals, and universities. In 2012 the city reevaluated the program and decided that declining private interest rates and increased policy attention from the provincial government to climate change meant that using reserve funds to facilitate energy conservation investments was no longer necessary. The remainder of the funds was reallocated to address problems from the emerald ash borer, which had been wreaking havoc on the city's trees.

In its place, the city established a low-interest loan program, called Sustainable Energy Plan Financing, with a capacity of around $160 million. The city loans money to building owners at its rate of borrowing and has aligned the program's priorities with the city's sustainable energy plan. The new program—dubbed the Better Buildings Partnership—provides low-interest loans to public, institutional, and, increasingly, private building owners interested in energy conservation retrofits. Borrowers must also demonstrate how they are leveraging the funds with other sources, meaning the city's investment can be multiplied several times over.[61]

What is particularly innovative about this approach is the fact that the loans are considered recoverable funds, and therefore they are not tied to the tax base or contribute to the city's debt limit—it is an "off-balance-sheet financing mechanism."[62] In fact, the vast majority of the Energy and Environment Division's programs do not rely on tax dollars, which allows the division to avoid having to gain approval from the city council, which has historically been unwilling to invest in programs with long pay-back periods.[63] Uptake of the program has primarily come from city agencies, city corporations, Business Improvement Areas, and other organizations affiliated with the city, though it is slowly expanding into the broader community, including universities and hospitals.

The city has also reduced the barriers to energy conservation by providing new sources of financing for single- and multifamily residential building owners. This differs from Los Angeles's approach, which has been to incentivize or even partially fund such measures directly. Toronto's Home Energy Loan Program (HELP) and High Rise Retrofit Support (HiRIS) programs provide homeowners and property owners with very-low-interest loans for energy conservation retrofits. The loans can be paid off over time on the owner's property tax bill; in this way the debt remains with the property rather than with the owner.

Championed by city council member Michael Layton and supported by TAF, these programs use the refundable debt model first developed through the Better Buildings Partnership. The city currently has a $20 million pilot program running, but the limit is somewhat arbitrary, as borrowing for the program does not count toward the city's overall debt.[64] The city recovers the administrative costs of the program with a small surcharge and charges interest at the city's rate of borrowing. The program was developed during Rob Ford's mayoralty, and at a time when the city was approaching its debt limit of 15 percent of carrying costs. The fact that it is revenue neutral helped significantly in gaining support from the city council.⁶⁵ The innovative programs were recognized with a 2016 Sustainable Cities Award from the Federation of Canadian Municipalities.

One important target of the HiRIS program in particular is the city's aging stock of multifamily housing. Hundreds of concrete slab apartment buildings were built in the city between the 1950s and 1980s and are highly inefficient energy users. As part of the Tower Renewal program originally developed by David Miller, which has a broader aim of revitalizing both the towers and their neighborhoods, the HiRIS program has provided a much-needed revenue stream for apartment building energy retrofits.

Reducing City Government Emissions

All three cities have taken steps to reduce emissions from city government operations. The cities recognize this as a useful strategy for demonstrating leadership and commitment, and it allows them to implement new programs and projects relatively independently. Toronto is distinct in that it has made reducing city government emissions a large focus of its work and, due to the relatively high contributions from the waste sector, reducing city government emissions can make a significant contribution to the city's overall mitigation goals. This focus has also come about in no small part because of the limited regulatory tools and political leadership the city has had for addressing climate change.

Toronto has targeted city buildings, the city's vehicle fleet, and landfills as opportunities to reduce GHG emissions. City council has passed legislation that requires a business case assessment for deep building retrofits whenever a building is being renovated. Funds from the Better Building Partnership have been used to retrofit dozens of city-owned buildings. The city has also partnered with Toronto Hydro to leverage the provincial feed-in tariff program and install solar panels on city-owned buildings. In 2014, the city council went as far as to pass legislation requiring that all new city buildings to generate at least 5 percent of their energy from renewable sources. Toronto has also significantly downsized the city's vehicle fleet, switching to lighter-utility trucks and lower-emission models.⁶⁶

The most ambitious project to date has been the city's investment in methane-capture technologies in the city's five landfills. Backed by seed money from TAF, the city has built a number of facilities able to transform methane from waste into useable fuel. A private company, Eastern Power Developers, was then selected to install a power plant at the site. The project at Toronto's primary landfill generates enough electricity to power 24,000 homes (C40 2012b). The city's waste division is currently developing a long-term strategy that centers on sustainability and reducing GHG emissions, and it includes a number of new ideas such as investing in renewable natural gas and installing solar panels on landfill sites.⁶⁷

Toronto is currently seeking to leverage its work to reduce city government emissions into carbon credits the city can sell. By 2012, Toronto had reduced the city government's own emissions by 49 percent compared to 1990 levels (well above the goal of 6 percent). Around this time, the city also realized that there was the potential for these innovations to generate additional revenue for the city if they were sold as carbon offset credits. There is an emerging private market for such credits, and nearby cities had had some success in selling credits to large companies interested in a "greener" portfolio.

This work has been driven by staff, and led by the city manager, who wanted greater diligence from staff in safeguarding emissions for a future with a market.[68] Through research, the city discovered that companies entering this voluntary market are interested in offsets that also allow them to make claims about social responsibility and giving back to the community. In a voluntary market, different tons of reductions are not necessarily equal—energy efficiency in libraries and civic centers, waste to energy schemes, and anaerobic digestion of compost come with strong social value. A company interested in investing in carbon offsets wants them to come with a story and guarantees the offsets are retired rather than bought and sold in a compliance market.

Toronto saw similar cities in the region making money in this voluntary market and creating capacities and protecting assets for when the province's compliance market eventually came into place. In 2013 the city council approved a carbon credit policy, "which outlines conditions under which the City will sell its carbon offset credits including a condition that credits, once sold, must be retired immediately and not re-sold to potential polluters" (City of Toronto n.d.). It also requires the emission reductions are in excess of the city's own emission targets, and it limits sellable carbon credits to 50,000 tonnes. Developing carbon credits helps validate the city's work—that they have produced real, quantifiable, and verifiable emissions savings.

The program has fostered a diligence in city government to be aware of and track all GHG emission savings to protect the city's ownership of these. To date the city has identified three carbon credit projects: a waste-to-energy project at a city landfill, energy efficiency in city-owned buildings, and a retrofit of Toronto Public Library buildings. By taking its success in reducing city GHG emissions and transforming this into an institutionalized carbon credit, the city has helped to ensure that the credits are retired and have a real impact on GHG emissions. The development of carbon credits has been a staff-driven initiative, with the city manager aiming to safeguard emission reductions for a future with a market.[69]

Providing Low-Emission Public Housing

While Toronto has primarily relied on enabling and self-governing modes of governing GHG emissions, it also has taken a particularly innovative approach to reducing emissions through retrofitting public housing buildings. In 2007, TAF developed a new tool for financing building energy retrofits called the Energy Savings Performance Standards. This financing tool uses a percentage of the verified energy bill savings to repay the initial investment in an energy efficiency project. Like the city's own programs, this is a nondebt instrument and typically would extend for a period of ten years. It also developed a strategy for undertaking deep building retrofits called Integrated Project Delivery, which aligns various elements and actors involved in a building retrofit around shared goals.[70]

Toronto's social housing buildings are prime targets for this type of program. The buildings are in a perpetual state of disrepair, and the Toronto Community Housing Commission (TCHC) faces persistent budget shortages and backlogs. In 2012 the Robert Cooke Co-op, a social housing complex with 28 townhomes and a 123-unit tower, expressed interest in using TAF's new financing mechanism for energy retrofits. TAF took it on as a way to pilot its new tool and aimed to reduce the building's GHG emissions by 30 percent, generate $65,000 savings in annual utility costs, and improve indoor air quality. The project ultimately exceeded expectations in every category, reducing GHG emissions by around 200 tons per year (Toronto Atmospheric Fund 2016).

Following this successful pilot, TAF approached the TCHC about the possibility of developing a more ambitious effort to reduce energy and water use in public housing buildings. TCHC houses around 164,000 residents in 2,200 buildings across the city (Toronto Atmospheric Fund 2012); for them, such an initiative represented a way to fund a backlog of building repairs and maintenance that were leading the agency to consider closing units in very bad repair. For TAF, the partnership represented an opportunity to support several benefits at once: housing affordability, improved living conditions, and reduced GHG emissions.

The two agencies launched the TowerWise program in 2014 with direction from the city council and a $3 million loan from the Federation of Canadian Municipalities. They chose to target seven buildings, with 1,200 households, for an initial round of deep building retrofits. Construction was completed around the end of 2016, and an extensive system for monitoring and verifying the savings and air quality improvements that result from the project is in place. In 2016 the provincial Ministry of Energy and Climate Change recognized the TowerWise program with an Award for Environmental Excellence.

A Transportation Vacuum

Policies for reducing GHG emissions from transportation have been conspicuously absent from Toronto's agenda, despite the sector's large contribution to the city's GHG emissions and growing frustration among Torontonians with the city's transportation system. Toronto's original climate change plan, *Change is in the Air*, outlined plans for *Transit City*, Mayor David Miller's vision for connecting the city through a network of light rail. It also highlighted the potential for expanding bike lanes, shifting taxis to hybrid vehicles, and working with the province to promote the city's goals. *Transit City* in particular represented an ambitious plan for expanding public transportation and increasing mobility in Toronto and was supported by a broad coalition in the city council that included both downtown and suburban representatives. The province was also behind the plan and had committed C$8.3 billion to the initiative, along with C$317 million from the federal government.

When Ford was elected in 2010, *Transit City* was wiped from the books. The suburban–downtown coalition Miller had built also dissolved.[71] Ford's goal for transportation was to make it easier for people to commute by car, which for him and his coalition meant prioritizing subways rather than surface modes of public transportation. "It was the most anti-transit, pro-transit argument I've ever heard."[72] Ford also tapped into suburban resentments of the downtown "elite" who had access to a well-serviced subway system, and framed light rail as a second-best option. He developed a subway plan for the city but the council refused to approve it, ultimately reaffirming its commitment to a light rail–based transit system. The city's inability to develop consensus on what it desired for transportation led to inaction at both the local and provincial levels.

Mayor Tory has made transportation a priority for his administration but rarely links transportation and GHG emissions. His *Smart Track* proposal for regional express rail lines and subway extensions was a central platform for his mayoral campaign. The primary challenge for the city has continued to be disagreements within the city council about the modes of transit that should be prioritized—who and what the transportation network is for. High-profile political battles over a subway extension to Scarborough, one of the city's least served areas, and the aging Gardiner Expressway have largely fallen along suburban–downtown lines. The resolutions have been compromise positions that do not often make economic sense, let alone maximize GHG emission reductions. For example, the city council recently approved a $3.3 billion subway extension to Scarborough that has one stop—Scarborough residents get their subway, but the city only has to pay the cost of a single station. Even a pilot bike lane along a small portion of Bloor Avenue had significant difficulty getting through the council following an eight-hour debate.

Toronto's fractious internal politics—a legacy of amalgamation—have stymied transportation policy despite (or perhaps because of) relatively high levels of local authority on the issue. The downtown–suburban divide also shapes provincial priorities. Mayor Tory's recent request to introduce a road toll on the Don Valley Parkway—a measure intended to reduce congestion and raise money for public transportation—was denied by the province despite earlier signals of support, largely due to electoral vulnerability in Toronto's suburban ridings (Benize 2017). While Tory has been able to secure compromise on some transportation projects and help them to move forward, the city still lacks a coherent vision of its transportation future and how it will further its efforts to reduce GHG emissions. In a city with nearly half of its emissions coming from transportation, this will be a critical obstacle to overcome in the years ahead.

Tailored Policy Agendas for Climate Change Mitigation

The policy agendas New York City, Los Angeles, and Toronto are developing to reduce GHG emissions reflect the unique contexts they confront: the variation in their sources of emissions and in the strengths and weaknesses they bring with them to confront the challenge. Each city has worked strategically to develop a set of policies and programs tailored to their conditions.

The cities have made different choices about which sources of GHG emissions to target, reflecting both the importance of these emission sources and the opportunities the cities have for taking action. New York City has invested significantly in reducing energy used in buildings, the largest source of emissions for the city but also an area where they have relatively high levels of authority. The city's ability to develop a city-scale energy code and general building codes has been an important opportunity for changing behaviors and reshaping the built environment.

Alternatively, Los Angeles has focused much of its effort on changing the city's energy supplies and expanding transit. This has had less to do with the city's GHG emissions profile and more to do with the fact that the city is able to work directly through the LADWP to determine energy investments and generation priorities and has enjoyed significant influence over transportation planning at the regional level.

Toronto has taken a less targeted approach, with a range of initiatives in the building sector and city government operations. Again this emerges from the city's relatively weak political and institutional position; city staff and committed councilors have had to take advantage of opportunities as they present themselves rather than pursue a well-charted implementation strategy.

The multiple governing modes available for reducing GHG emissions have allowed these very different cities to pursue their climate change mitigation goals in unique ways (table 3.2). The cities have worked strategically to take advantage of political and institutional opportunities for policy development.

New York City has relied heavily on its strong political leadership and hefty institutional powers to pursue a largely regulatory means of reducing GHG emissions. It has set rules and standards regarding how buildings should be constructed and operated, what kinds of fuels can be burned, and what building owners are required to share with the city. Los Angeles has taken advantage of its role in energy and transportation service delivery to provide these services in ways that produce fewer GHG emissions. The city's energy portfolio and transportation systems stand to be fundamentally transformed. In Toronto, climate change mitigation policy has often been led by the city's public service, who have found creative ways to enable others to take action and take full advantage of opportunities to self-govern. The creative work of city staff, and innovative spark from TAF, help demonstrate how a city without some of the resources of New York City and Los Angeles can still govern climate change mitigation.

The cities' experiences also demonstrate that climate change policy agendas blend and combine governing modes (table 3.3). For example, when New York City encountered obstacles to regulating emissions from cars it adopted a more provisioning-oriented mode of governing, focusing on reorienting transportation services and opportunities in order to reduce emissions. Conversely, Los Angeles has begun to use a more regulatory mode of governing building energy use, following New York City's lead to require energy use benchmarking and audits. Opportunistic Toronto has also used a number of governing modes to target building energy use.

Actors in different parts of city government are responsible for forwarding their city's climate change mitigation agendas. While in New York City the mayor has clearly been a central and very public figure for mobilizing resources and actors, in Toronto this work has largely been done quietly by city staff. In Los Angeles, policy has typically been conditioned by compromise between the positions and priorities of the mayor, city council, and the LADWP. A city's larger context also influences *who* within city government is developing the climate change mitigation policy agenda.

The ability to govern GHG emissions through multiple modes has also allowed the cities to learn and strategically adjust their approach over time. In each case, city government turned quite purposefully first to its strength: regulating energy use, changing energy services, or reducing barriers for others to act. Over time, the agenda expanded and diversified and the cities have moved from what might be viewed as "low-hanging fruit" to taking on more challenging emission sources

TABLE 3.2 Primary routes taken to reducing GHG emissions in New York City, Los Angeles, and Toronto

	EMISSIONS		GOVERNING MODE	
	SOURCE CHOICE	INFLUENCED BY	MODE CHOICE	INFLUENCED BY
New York City	Energy use in buildings	Main contributor of GHG emissions Use of formal authorities	Regulatory	Presence of formal authorities Collaboration between mayor and city council
Los Angeles	Energy generation and supplies; transportation	Use of formal authorities Opportunities for co-benefits (transportation)	Provisioning	Presence of formal authorities Opportunities for leadership
Toronto	Energy use in buildings	Use of enabling capacities	Enabling	Lack of formal authorities Innovation and capacities of city staff

TABLE 3.3 Use of governing modes in New York City, Los Angeles, and Toronto

	REGULATING	PROVISIONING	ENABLING	SELF-GOVERNING
New York City	High	Low	Medium	High
Los Angeles	Low	High	High	High
Toronto	Low	Medium	High	High

or political hurdles. The cities have had to learn from their mistakes and try things in a different way.

The Role of Context in Shaping Urban Climate Policy Agendas

The evidence from New York City, Los Angeles, and Toronto—cities in very different political, institutional, economic, and technical settings—demonstrates that context is more likely to shape *how*, rather than *whether*, these cities develop a policy agenda for responding to climate change. The differences between these cities—such as a weak mayor or a municipal utility—haven't deterred them from taking action, or even led to different levels of ambition. These differences do influence the choices they make about the sources of emissions to target and the modes of governing by which to pursue change. Each city has its own very local reasons for taking on climate change, and the policy agenda has emerged in different ways. In short, the policy agenda for governing climate change mitigation looks very different in different cities, even though the aim of significantly reducing GHG emissions may be the same.

There are several findings worth drawing out more explicitly. First, there is no single dimension of a city's context that itself determines the policy agenda it develops for climate change mitigation. The agenda is shaped by what Oran Young refers to as "interactive causal clusters," rather than simple causal chains of relationships or events (Young 2008, 10). New York City's agenda has been shaped by the opportunities presented by political leadership, its sources of authority, and a receptive and well-organized private sector. Los Angeles has tailored its policy agenda to its central role in service provision, a progressive state government, and local political considerations and tradeoffs. Toronto's agenda has been shaped by the legacy of institutional and financial investments in TAF, a dedicated public service, internal political divisions, and its relative lack of authority over GHG emissions. In each case these factors have worked together to shape the choices the cities make about how to pursue their GHG emission reductions goals but have not led to variation in their ambitions.

Critically, the variation in cities' policy agendas for reducing GHG emissions generates variation in the implications of climate change mitigation for the city more broadly. For example, significantly reducing GHG emissions in New York City may ultimately require very little substantive change in the way people move throughout the city. Instead, buildings and energy generation will change significantly. Alternatively, transportation systems in Los Angeles and Toronto will need to function much differently in the future if the cities are to reduce GHG emissions by 80 percent. Building owners in different cities will be faced with a unique mix of regulations and incentives, and residents might be asked to adopt more or less personal responsibility in their choices and behaviors for meeting the city's targets. Climate change mitigation policy agendas have different effects on different cities and relate in different ways to the lives of urban residents.

In the three cases presented here, city governments are taking advantage of and leveraging federal and state/provincial programs rather than being driven by them. Opportunities for funding or incentives for action in a certain area (e.g., Ontario's feed-in tariff system and coal phase out) might provide cities with a boost or a nudge, but they are rarely the driving force behind the city's climate change work more broadly. Rather, local political dynamics, opportunities for leadership and co-benefits, and commitment from decision makers and staff are what serve to initiate and sustain local climate change action.

Even in Los Angeles, where the state of California has been extremely progressive on climate change, the city has consistently sought to stay ahead of the state and maintain a leadership position. Governor Brown saw state-level policies as "the tip of the spear, the bleeding edge,"[73] and has shown that California is reducing GHG emissions and growing the economy at the same time, creating jobs during an economic downturn. But Los Angeles makes a point of explicitly going beyond the state's targets. Federal and subnational frameworks shape a city's context and its policy agenda, but ultimately local factors and governing efforts are what drive and sustain action on climate change mitigation. This finding aligns with scholarship on other dimensions of urban policy, demonstrating that local political and institutional conditions shape the variety of responses of cities to trends or priorities at higher levels of government (Clarke and Gaile 1998; Sellers 2002a).

Finally, the unique and dynamic relationship between a city government's climate change mitigation policy agenda and its context has implications for the way we draw lessons from these experiences for other cities. Recommending Los Angeles's solar power energy programs to a city without a municipally owned energy utility will likely not be very useful. Similarly, understanding the changes to the building code adopted by New York City will not be of much use to cities without a code of their own. In order to effectively learn from cities' experiences with climate change mitigation, it is paramount to understand the broader context in which they were developed and the conditions under which a similar program or policy might be feasible.

More likely to be transferrable between cities are the governing strategies used for developing and implementing new climate change policies, maintaining support for climate change, and assembling the resources needed to make progress. The *process* of developing a policy or program is potentially more valuable to urban decision makers and advocates than the details of the policies themselves. Identifying and understanding these governing strategies also allows us to dig deeper into the means by which city governments are able to mobilize actors and resources toward the new and transformative aim of responding to climate change. In the next chapter I take up the task of examining such shared governing strategies and specifically the role institution building, coalition building, and capacity building have played in facilitating urban climate change mitigation.

4
THE MEANS BEHIND THE METHODS
Governing Strategies to Reduce Greenhouse Gas Emissions

New York City, Los Angeles, and Toronto have developed climate change mitigation policy agendas that rely on a range of governing modes and target different sources of emissions, according to their particular opportunities and obstacles. They have each developed a tailored and evolving route to climate change mitigation, despite differences in capacity, resource availability, policy context, and formal authority. Rather than influencing their choices about climate change commitments, these contextual factors have shaped the cities' choices about how to pursue their goals.

Documenting the choices the cities have made about their policy agenda for reducing GHG emissions and their relationship to context provides just half of the story. It helps us to understand the routes cities take but not the means by which they are propelled forward. Steering a city in a new direction requires the reconfiguration of the political, institutional, and organizational landscape of decision making, or "crafting arrangements through which resources can be mobilized, thus enabling a community to accomplish difficult and nonroutine goals" (Stone 1993, 1). Climate change mitigation is indeed a "nonroutine goal" for the majority of cities. Therefore, understanding urban governance for climate change mitigation also requires examining the *strategies* city governments use to overcome this uncertainty, craft new arrangements, and fuel the development of programs and policies for reducing GHG emissions.

As David Jacot at the Los Angeles Department of Water and Power points out, "what is hard is going from agreeing on the big picture to doing the little things that will get you there . . . you have to go step by step, brick by brick of the strategy."[1] In chapter 2, I identified three governance strategies available to, and required of, city governments for climate change mitigation: institution building, coalition building, and capacity building. Each strategy reduces key sources of uncertainty that arise when taking up the new and challenging issue of climate change mitigation. The governing strategies facilitate action on climate change and channel resources to the effort. In this way the strategies underpin and support governance for climate change mitigation regardless of the particular mode of governing or source of emissions being targeted.

In this chapter I examine how New York City, Los Angeles, and Toronto have used these governing strategies to support their efforts to reduce GHG emissions. Their experiences demonstrate the centrality of the three governing strategies to mobilizing the resources and actors needed to reduce GHG emissions. The city governments are adapting and adjusting their governing strategies over time as they learn more about the challenges and uncertainties that come with reducing GHG emissions. The cities are at different stages in their use of the three governing strategies and have at times placed more focus on some than on others. It is clear that they recognize these strategies as central to their work on climate change mitigation and use them to facilitate their efforts to steer their cities in a new direction.

Institution Building

The institutions of city government are not always well-suited to the task of reducing GHG emissions (Aylett 2013; Hughes 2017). Climate change mitigation introduces a new goal for city governments—one that may not fit easily within existing administrative structures or budgetary processes. The work of reducing GHG emissions makes new demands on existing decision-making structures and requires new means of collaboration and coordination across sectors and departments. As a result, governing climate change mitigation engages with significant institutional uncertainty about who is responsible for the work and how this responsibility will intersect with existing practices and administrative structures. This uncertainty must be overcome to mobilize the needed resources and actors. Institution building helps cities overcome this uncertainty by building up the "resources for change embodied in the decision-making procedures and rules for the city" (Romero-Lankao, Hughes et al. 2013, 787). The experiences of New York City, Los Angeles, and Toronto reveal two particular ways city governments use institution building to reduce uncertainty: institution building for coordination and institution building for longevity.

Institution Building for Coordination

The cities have developed institutional strategies for facilitating the coordination needed to develop and implement climate change mitigation policies and programs. This has typically involved starting internally with coordination between departments and between elected officials and administrators. Because climate mitigation commitments have been mayoral initiatives, each administration has had to make choices about where and how to administer climate change mitigation programs as they do not fall neatly along administrative lines. They have sought to carve out explicit institutional space for climate change mitigation and give structure and guidance to those taking responsibility, both within and beyond city government.

Working closely with the city council, Mayor Bloomberg in New York City made explicit use of internal institution building to coordinate the city's efforts to reduce GHG emissions. Rohit Aggarwala, head of the Office of Long Term Planning and Sustainability (OLTPS) under Mayor Bloomberg and a former management consultant, refers specifically to the city's strategy as one of "change management," noting that the city's work on climate change "was not just a policy effort."[2] The city recognized the institutional uncertainty of climate change mitigation, noting the core challenge was that "we have commitment from the top but uncertainty about how to get it done, how to get everyone on board."[3] The mayor's staff thought explicitly about how to reorient institutional structures in a way that would facilitate progress on their climate change agenda.

Soon after the release of *PlaNYC*, the mayor and city council both went to work to ensure sustainability and climate change were embedded in city decision-making frameworks. Mayor Bloomberg created the Office of Long Term Planning and Sustainability (OLTPS) within his administration for this purpose and tasked it with leading implementation of *PlaNYC* and serving as a bridge among departments such as Environmental Protection, Sanitation, Transportation, and Parks and Recreation. Importantly, the OLTPS was also relatively well staffed, initially with twenty people.[4] The creation of the OLTPS represents new commitment at the mayoral level to the issue of sustainability. It provided an additional signal from the mayor that sustainability and climate change were priorities for his administration. The OLTPS became an in-house think tank for the city and a hub for interagency coordination. Nongovernmental organizations (NGOs) and advocates with an interest in the city's mitigation efforts also saw the OLTPS as providing a useful a one-stop shop for accessing the administration. Working in the OLTPS was considered prestigious: it was easier to make the case for implementing something if it was linked to *PlaNYC*.[5]

The OLTPS played three different roles in reducing institutional uncertainty.[6] For some measures in *PlaNYC*, there was a clear lead agency that was eager and able to move forward. For example, the Transportation Department had a clearly defined set of measures in *PlaNYC* and a commissioner, Janette Sadik-Khan, who was on board with the mayor's plans. In such a case, the OLTPS would play a support role if the agency needed help with budgeting, capital, or the state legislature. For other measures, there was a clear lead agency but it needed more support, either because they weren't interested, didn't have the staff, or were required to do a big "political lift" to meet the aims of *PlaNYC*. In this case the OLTPS was likely to take on themselves the portions of the work the agency was unable or unwilling to do. For example, the Department of Buildings was understaffed, so the OLTPS put together a team of its own tasked to work with the department on energy efficiency measures. Finally, there were measures in *PlaNYC* for which there was no existing lead agency. Adapting to climate change was a clear example of this type, and the OLTPS took this work on themselves. In each case the administration actively fostered and strengthened what they called "leadership that compels people to do things and not have to worry about whether they will or not."[7]

Getting city agencies on board with a mayoral agenda is challenging, even when the mayor is powerful and a strong leader, as was the case with Bloomberg. "Agencies are notoriously resistant, and require internal advocates . . . this creates a 'native' constituency for a sustainability agenda and makes it possible for ideas to be originated internally."[8] Using the OLTPS as a central hub for implementing *PlaNYC* relied on a top-down approach to implementation that received a mixed reception within the city's agencies. In some cases Bloomberg appointed agency heads more likely to respond to his style. Ultimately, many city agencies institutionalized *PlaNYC* through the establishment of agency sustainability divisions and the appointment of senior staff members responsible for overseeing *PlaNYC*'s initiatives (City of New York 2013). In many cases the aims of *PlaNYC* became written into agencies' goals and plans. For example, in 2008 the Department of Transportation released its *Complete Streets* plan, which included strategies for reducing the department's GHG emissions in order to build toward the city's larger goals (New York Department of Transportation 2008).

In Los Angeles, both Villaraigosa and Garcetti have relied on the mayoral power of administrative appointments to facilitate coordination on climate change mitigation. Both mayors worked actively to identify champions in key departments, and Garcetti has gone so far as to state explicitly that reviews of general managers will include whether they are meeting the goals of the city's climate change plan (City of Los Angeles 2015).

Building a larger institutional structure for coordinating climate change work in the city has been more of a challenge for Los Angeles. When *Green LA* was released in 2007, accountability and coordination for the implementation process was to be provided by a new entity established by Villaraigosa within the city's larger Environmental Affairs Department (EAD), Environment LA. Environment LA was tasked with providing annual GHG emission inventories, and departments were to submit regular progress reports to them. This arrangement was very short-lived. The city's myriad departments and agencies balked at requests for GHG emission data, fearing political or administrative repercussions if found to be egregious emitters. There were also political tensions between Villaraigosa and the city council that worked to prevent council buy-in to the *Green LA* plan.

These tensions came to a head following the 2008 U.S. housing recession, which had particularly devastating effects for Los Angeles: median income in Los Angeles County fell 11 percent, home prices fell 45 percent, unemployment was below the national average, and 52 percent of the area's families were considered low income by 2010 (Bohn and Schiff 2011). City staff experienced layoffs and furlough days, and support for the *Green LA* initiative lagged. The city council wanted a high-profile cut and voted to dismantle the city's Environmental Affairs Department. The move was led by Jan Perry, long-time chair of the Energy and Environment Committee, and seconded by future mayor Eric Garcetti.

This department had been a city council target in the past, but it had always managed to stave off attacks. Some councilors claimed that integrating the city's work on environmental protection and compliance within existing departments was a more effective strategy. This time, fearing a veto override, Villaraigosa decided to develop a plan for getting the work done without the EAD. He bolstered his office staff, borrowed people from elsewhere in the city, and worked more closely with the Department of Sanitation and the LADWP. Work on climate change mitigation was shifted to the Department of Sanitation, but many stakeholders did not perceive Sanitation as a neutral, professional group the way the Environmental Affairs Department was.[9] The task of conducting a GHG emission inventory was also given to the Department of Sanitation, but policy implementation tracking was virtually nonexistent and any reductions did not feed back into the inventory as planned.[10] In fact, a new city-wide inventory would not be completed (and publicly released) until 2016, and tracking progress on *Green LA*'s goals was almost entirely absent. Garcetti has moved oversight and coordination of his *pLAn* to the Mayor's Office of Sustainability, a small unit with around four staff members.

A more innovative example of institution building for coordination in Los Angeles comes from the LADWP. In Los Angeles, electricity and gas are managed separately, though both make significant contributions to the city's GHG emissions profile (32 percent and 9 percent, respectively). LADWP has responsibility for electricity while the Southern California Gas Company (SCG), a private utility, has responsibility for gas. Both utilities have conservation and efficiency goals to meet.

Managers at the LADWP realized the separation in service created barriers to innovative conservation and GHG emission reduction projects and generated redundancy and overlap in programmatic offerings. Undertaking joint projects required negotiating responsibility and financial contributions on an individual basis. To solve the problem, the two utilities developed a "master inter-utility agreement" specifically designed to foster greater collaboration on energy and water efficiency. As a manager at LADWP described it, "the lawyer stuff is handled in one large master agreement, a set of general rules, and then the programs all adhere to them. It outlines who does what and who pays for what . . . we can still claim credit for the things that are most important to us for regulators, customers, etc."[11] There are currently thirteen joint programs in place through the inter-utility agreement for both energy and water efficiency. Institutionalizing cooperation cleared the way for innovative energy and water efficiency solutions.

In Toronto's case the mayor lacks many of the powers of appointment enjoyed by the other mayors, and the city's agencies report to the city council. To build institutions for coordination, therefore, Mayor David Miller worked to signal that climate change was a priority for him and for the city in a variety of ways. He invested significant political capital in securing unanimous approval from the city council of the climate change targets and measures included in *Change is in the Air*. "It showed it was not just a mayoral thing, city council support is a big signal to the public and to the administration . . . I used everything possible in my arsenal of persuasion: telling people this is important to the mayor, he'll remember it, etc."[12] To outsiders, the council's vote was a "clear signal to the bureaucracy and Torontonians that the mayor cares about climate change."[13]

Despite council approval, the city still faced the challenge of finding an administrative home for climate change mitigation. As in most cities, even when climate change legislation passes through the city council, there is still work to be done to make it a priority for the civil service.[14] Miller initially coordinated oversight from his office, working to find people in departments who could serve as champions. Under pressure from competing progressive claims from his coalition, Miller was not able to prioritize climate change mitigation through the end of his term. The Energy and Environment Division has since become the lead on implementation and further policy and planning development for reducing GHG emissions, but it does not have the ability to initiate projects that fall outside of its mandate.[15] The city manager has not shown interest until recently, nor has he been directed to do so by city council. Toronto still struggles with the question of how to coordinate the city's internal work on climate change mitigation.

The Atmospheric Fund, or TAF, is one institutional mechanism Toronto has for facilitating coordination on climate change mitigation. Indeed, part of TAF's mission is to serve as a bridging institution in the city, helping to coordinate between city departments and between the city and its stakeholders. TAF has the advantage of operating outside of the institutional siloes and bureaucratic rules, political considerations, and decision-making procedures that can stymie coordination within the city.[16] TAF looks for climate change leaders in the community and helps to mobilize them and connect them with the city. They also identify and help to fill important gaps that are preventing coordination, and especially GHG reduction opportunities that are slowed down by a lack of funding. For example, TAF has led pilot programs to install LED street lighting and retrofit social housing buildings, has developed new financial tools for energy conservation retrofits in commercial buildings, and helped mobilize diverse stakeholders around sustainable transportation issues in the 2014 mayoral election. Establishing TAF in 1991 and including coordination as part of its aim is an institution-building investment that has continued to pay off for Toronto, particularly as other efforts to provide this coordination internally have stalled.

Institution Building for Longevity

A second way of approaching institution building has been initiating changes that ensure the longevity of climate change mitigation initiatives. Reducing GHG emissions is a long-term project, and the goals these cities set for themselves extend beyond the tenure of any single mayor. As climate change mitigation represents a new direction for the cities, institutional changes are required to ensure that resources are directed to these programs even after a mayoral transition. Mayors and councilors in New York City, Los Angeles, and Toronto have been aware of this challenge and have taken steps to build institutional arrangements able to insulate climate change mitigation from political shifts and incorporate, if not prioritize, climate change mitigation in the long term.

Here again New York City has been the most deliberate and far-reaching through its use of a change management strategy. Both the mayor and city council "thought a lot about how to down cycle, to insulate from change."[17] The targets and programs laid out in *PlaNYC* are not themselves legally binding in any way. Both the mayor and city council were concerned about the consequences of a new mayoral administration and wanted to ensure that the city's climate change goals were more formally codified. The city council was also interested in having more input into climate change priorities and plans for the city going forward.

Speaker Quinn and James Gennaro, then chair of the council's environmental committee, sought to hard-wire *PlaNYC* and its GHG emission reduction goals into the city. The city council unanimously passed two key pieces of legislation. The council required the city maintain both a GHG emissions reduction target and a plan for meeting the target through Local Law 55 of 2007 (the New York City Climate Protection Act) and Local Law 17 of 2008. The legislation also codifies the aims of *PlaNYC* and the city's GHG emission reduction targets (30 percent by 2030 citywide, 30 percent by 2017 for city operations). The council refers to the targets as a "soft mandate," not a goal, because the legislation stipulates that if a mayoral administration cannot meet the targets it must notify the city council and provide recommendations for how to move forward. The legislation requires an annual inventory of citywide GHG emissions, a description of the city's efforts to reduce emissions over the past year, and the continued use of a Sustainability Advisory Board comprised of representatives from environmental, environmental justice, planning, labor, business, and academic sectors. The city council also codified the mayor's use of an office of long-term planning and sustainability and charged this office with developing and implementing a long-term sustainability plan for the city that included climate change.

By the time Bloomberg left office, climate change mitigation was hard-wired into the city. Importantly, the legislation included flexibility for each mayor to carve out his or her own approach to climate change and integrate it with their particular agenda. The legislation doesn't specify particular outcomes or objectives beyond reducing GHG emissions and requires the mayor develop a plan toward these ends.

When de Blasio was elected in 2013, he did not show much interest in environmental or climate change issues; his passion was tackling income inequality and poverty. De Blasio also needed a way to distinguish himself from his predecessor, whose legacy was embodied in *PlaNYC*. The new city council speaker, Melissa Mark-Viverito, and chair of the Environmental Committee, Costa Constantinides, were both committed to climate change and the aims of *PlaNYC*. After *PlaNYC* there was an expectation for New York City to lead on climate change, and they worried de Blasio's interest in income inequality would take the focus away from the environment and having NYC be an "environmentally premier city."[18] In 2014 one of the first things the new city council speaker did was demonstrate climate change and sustainability were still a priority for the council—and should be for the incoming mayor—by unanimously passing Local Law 66, which established a more aggressive GHG emissions reduction target of 80 percent below 2005 levels by 2050.[19] Therefore, rather than dissipating, New York City's climate change ambitions only intensified after Bloomberg left office.

In Toronto, codification has also been important for ensuring the longevity of climate change mitigation policies. When Rob Ford was elected in 2010, mayoral leadership on climate change ended for Toronto. The city council had previously approved the climate change targets and programs in *Change is in the Air*, and they remained a mandate for city agencies. This allowed staff to continue their work to reduce GHG emissions through the Ford administration, absent any mayoral leadership on the issue. Many city staff, especially those in the Energy and Environment Division, are deeply committed to climate change mitigation and have honored that mandate.[20] These same staff members have often sought to maintain a low political profile, and resources from city council remain scarce.

Los Angeles's efforts to codify climate change planning and targets have been stymied by tensions between the mayor and city council, particularly during Villariagosa's tenure. The city council did not adopt Villaraigosa's climate change plans or fully buy in to the goals these plans laid out. Groups peripheral to the *Green LA* coalition perceived the process as opaque and criticized the plan as lacking specificity. Indeed, the implementation strategy in *Climate LA* remained short on details and was not informed by a larger public conversation or consultation.[21] While *Green LA* had significant support from its core coalition, it was not universally embraced and failed to find a lasting foothold. Some councilors chafed at Villaraigosa's aggressive political style and choice of tactics, and the mayor had to largely rely on executive orders and political appointments to forward his climate change agenda.

These dynamics are beginning to shift, as council support and demand for climate change action grows. In 2014, the city council and the Board of Water and Power Commissioners explicitly went beyond this target and adopted a goal for the city of reducing energy use 15 percent by 2020. In 2016 Los Angeles city council committed to the GHG emission reduction targets laid out in the *pLAn* and instructed the LADWP to "study an equitable transition away from fossil fuels to powering Los Angeles with 100 percent clean energy" (Dhanak and Levy 2016).

The work that has continued in the city through and since Villaraigosa's initiative has largely been channeled through the LADWP and their ongoing targets and goals for energy services. The Board of Water and Power Commissioners under Villaraigosa committed to fulfilling at least 20 percent of the city's energy needs through renewable sources by 2017. The Board had also committed to achieving energy efficiency savings of at least 10 percent between 2010 and 2020. These mandates served as guideposts for the utility regardless of what was happening in City Hall.

Measure R, passed by voters in 2008, has served much the same purpose for transportation. Measure R ensured ongoing funding for expanding public transportation and a slate of projects designed to reduce GHG emissions. Environmental organizations also remain an important political force in Los Angeles, and during the 2014 mayoral election both major candidates had a strong environmental platform and sought out environmental endorsements from a community eager for action. As a result, Garcetti had an environmental mandate from the city coming in to office. Institution building for longevity can therefore take multiple forms, but it is critical to ongoing efforts to reduce urban GHG emissions.

Coalition Building

Significantly reducing urban GHG emissions requires the buy-in and support of a range of actors in the city and beyond. Private building owners, NGOs, voters, and professional associations are all key actors needed for the city to move forward with mitigation plans. The complex network of actors at play creates strategic uncertainty in efforts to govern the city for climate change mitigation. Different actors have different interests at stake; some may not be sure what their interests even are in this realm.

Coalition building is therefore a valuable governance strategy for overcoming this strategic uncertainty, serving to mobilize the political support and resources necessary for the development and implementation of new climate change policies and programs. Coalition building is a second way city governments can activate their capacity to govern climate change mitigation by building support among the city's various stakeholders and coordinating the range of political and financial resources needed for change. The experiences of New York City, Los Angeles, and Toronto reveal two different types of coalition building that mobilize different actors and resources for climate change mitigation: building stakeholder coalitions and mobilizing voter support.

Stakeholder Coalitions

In pursuing their climate change mitigation goals all three cities have sought to build diverse stakeholder coalitions that support new programs and initiatives. These efforts have often targeted some combination of environmental, social, business, and labor organizations in the city. Generating buy-in for city climate change programs helps to mobilize political will and resources and facilitates implementation (Mathur, Price, and Austin 2008; Anguelovski and Carmin 2011). Building stakeholder coalitions has also allowed the cities to tap into the expertise these groups hold in some of the technical aspects of climate change mitigation programs, such as building design specifications for energy efficiency, also helping to reduce substantive uncertainty.

New York City has built stakeholder coalitions through the use of stakeholder advisory groups that provide policy recommendations to the city. One prominent example is the Green Codes Task Force, which the city used to guide the process of greening the city's building codes to facilitate emission reductions. The city led a collaborative process to help overcome the challenges of conflicting viewpoints and to build technical expertise into the process. They also leveraged outside funding from private foundations, law firms, and the state government to support the task force's efforts.

The choice to use the Green Codes Task Force and seek out support from the private sector was highly strategic. While the city clearly had the authority and political will to change municipal codes, they were cognizant of the fact that developing appropriate legislation and ensuring compliance with the new rules would require significant buy-in and contributions from stakeholders. The city thought it was important that they were not seen to be driving the process and felt they lacked the technical capacity to do so regardless.[22] Revising building codes is a very technical exercise that requires expertise in building design and energy conservation technologies.

The real estate industry was a particularly important stakeholder in this process. The city knew it couldn't get climate change legislation through if it was opposed by the entire real estate industry.[23] They had to win over a segment of the real estate community and negotiate with the Real Estate Board of New York. The city therefore "outsourced" the task they referred to as "greening the codes" to the task force. The legislation that resulted from the recommendations of the task force, called the "green bills," were slated to go through the Housing and Buildings Committee in the city council, which has a notoriously full agenda. As the bills were a product of a collaborative process that often started before the bills were even drafted, and were only introduced if the task force could reach agreement, they passed through city council with stakeholder support, and none were vetoed. Hours were spent negotiating each line of every bill, in part to guarantee support from the real estate industry. A legislative analyst for the city council called it "one of the most collaborative processes I've ever been a part of."[24]

Equally importantly, the task force's involvement in the policy-making process created a coalition of invested stakeholders and decision makers around the city's climate change goals and programs. The task force was convened by an organization (Urban Green Council) that was seen to be relatively politically neutral and technically competent, with one foot in industry and one foot in policy-making. The process was seen as credible and legitimate: at every step along the way industry was included as true partners and has subsequently been supportive. A "shared ideology" was developed among the city council speaker, mayor, environmental groups, and the real estate industry.[25] The city has generated significant buy-in from the real estate and development industries to the larger project of GHG emission reductions, such that they are now considered an ally in these efforts rather than a source of political pushback (as they seemed poised to be initially). When the city introduced a target of reducing GHG emissions 80 percent by 2050, stakeholders bought in and "now everyone knows each other, we are smarter, and have created a body of knowledge. We have connected a community, which is maybe the most important thing we have done and it wasn't on purpose."[26]

Along with the codification of the city's climate change targets, this engaged stakeholder community also motivated de Blasio's commitment to climate change mitigation. Stakeholders and city staff became committed to and engaged with climate change mitigation and expected any new mayor to continue to make progress on the issue. When nine months had passed without de Blasio appointing a new director for the OLTPS, a group of stakeholders including the Regional Planning agency and the NRDC joined forces to convince the incoming administration that de Blasio could address climate change without compromising his interests in income inequality. They wrote an op-ed urging the mayor to not abandon the city's progress on climate change.[27] The mayor was also facing an upcoming UN meeting and climate week (including the People's Climate March) being held in New York City and risked embarrassment if the leading city on climate change did not have a plan in place going forward.

On September 21, 2014, on the eve of the People's Climate March, the de Blasio administration released *One City Built to Last* and announced its commitment to reduce emissions 80 percent below 2005 levels by 2050. In his statements, de Blasio tied this goal to his commitment to reduce income inequality and improve housing affordability, distinguishing his efforts from those of Bloomberg's. De Blasio positioned energy conservation as a tool for reducing energy costs for low-income people and made city support for conservation measures conditional on the preservation of affordable housing units. While he recognized the successes under Bloomberg, he argued that the 30 percent by 2030 goal was inadequate and introduced more aggressive targets of his own.

In Los Angeles, in addition to a municipal utility and a dedicated mayor, coalition building among labor, environmental, and business groups has been essential to its efforts to transform its energy system. The need for a supportive stakeholder coalition has driven the city's choice to prioritize local solar energy projects over more distant sources. The CLEAN LA Solar Coalition—which includes the Chamber of Commerce, neighborhood councils, real estate organizations, faith-based groups, the building community, and environmental groups—the Los Angeles Business Council, and the Sierra Club have been important supporters of local solar energy projects and increasingly involved in designing solar energy projects.

The LADWP has been working with these groups, through the mayor's office and with city council members, to jointly set targets and prices for solar energy through the feed-in tariff system. These prices (approved by the Board of Water and Power commissioners) have been negotiated to be higher than what LADWP standards alone would have produced, because they account for the avoided cost of transmitting energy and the economic benefits of developing local solar energy.[28] The LA Business Council was particularly interested in competitive prices for solar in order to attract solar development companies to Los Angeles and show that the LADWP is committed to the program.[29]

The LADWP has also learned to frame energy efficiency initiatives as good for business and jobs as a way of building support from key stakeholder groups in the city. As a senior LADWP manager put it: "translate the programs to jobs, tell the story, get the numbers, and shout them to the mountaintops."[30] This framing is thought to be particularly important when the city moves from setting an efficiency target to giving money to the particular projects needed to meet the goals. Energy efficiency programs use more labor and less capital than developing new sources of energy, including natural gas, and the jobs that are produced are more likely to be local. As a result, business groups in the city, such as the LA Business Council and Chamber of Commerce, as well as workforce development and tradespeople, have all been supportive of the city's efforts to improve energy efficiency.

Los Angeles also relied on stakeholder engagement for the development of its recent Existing Building Energy and Water Efficiency ordinance, which mirrors New York City's benchmarking approach. At the suggestion of a former Bloomberg staff member hired by the City of Los Angeles, the city undertook a nine-month engagement process to inform the development of the ordinance. The aim was to generate buy-in and understanding of the program and the advantages that businesses can gain from energy efficiency, such as cost savings and more valuable leasing opportunities.[31] This process was key to helping the ordinance pass through city council.

Toronto similarly found coalition building to be key for passing important local legislation for climate change mitigation. Toronto's Home Energy Loan Program (HELP) provides homeowners with very-low-interest loans for home retrofits that conserve energy. Councilors sponsoring the program presented it as revenue neutral: it would help people who couldn't otherwise afford energy efficiency upgrades, save taxpayers money, and create jobs.[32] Councilors and stakeholders met for one year with a volunteer consultant with knowledge of the U.S. Property-Assessed Clean Energy Programs.

The initial bylaw had the written support of a coalition of labor unions, environmental groups, civic organizations (Civic Action, Toronto Board of Trade), the Toronto Real Estate Board, and energy utilities (Hydro and EnBridge). Mayor Ford, not known to be a friend to environmental programs, supported the bill because of the large and influential coalition of supporters and the fact that it saved taxpayers money.[33] It went through the Economic Development Committee (rather than the Parks and Environment Committee) and was framed as a job creation measure in city council, further expanding its appeal for Toronto's diverse and nonpartisan city councilors. It was passed unanimously by the city council.

Toronto does not have the political commitment to climate change being shown by New York City and Los Angeles. Those in City Hall who are committed view coalition building as an important strategy for moving policy forward. One councilor says, "those of us who care have been careful to build the infrastructure outside of government and in the administration so that work is being done across the divides: public servants who can make the case for certain types of actions, and a constituency who is leaning on them for action."[34] Building a broader coalition of committed stakeholders helps the city govern GHG emissions absent strong political leadership at City Hall by mobilizing resources and actors for change.

Voter Support

The second way cities have used coalition building to fuel their mitigation efforts has been to mobilize voter support for taking the steps necessary to reduce emissions. This has meant mobilizing direct support for public referenda or ballot initiatives that require a majority (or super-majority) of voter approval, as well as public support for new policy initiatives.

This kind of coalition building has been central to Los Angeles's governing efforts. With the introduction of new energy conservation and renewable energy supply goals under Mayor Villaraigosa, it quickly became clear that the LADWP required an energy rate increase to fund the new programs. The utility is funded entirely through user fees and revenue bonds to raise capital funds, but it had not had an electricity rate increase since 1992. The LADWP has consistently charged its customers around 15 percent less on average than comparable (but privately owned) utilities such as Southern California Edison and Pacific Gas and Electric (California Energy Commission 2007). Unlike their private counterparts that are regulated by the California Energy Commission, increasing electricity rates for the LADWP requires approval from the Los Angeles City Council.

Far from a routine administrative exercise, raising electricity rates has proven to be a political flashpoint for the city, pitting residents against organized labor and testing the durability of the emerging labor–environment coalition behind the city's climate change initiatives. While a central motivation for electricity rate increases has been financing the programs and infrastructure necessary to reduce the use of fossil fuels and increase energy conservation, securing these rate increases has required an alignment of environmental, labor, and social justice advocates; mechanisms for differentiating between the needs of the central city and the more suburban areas; and a complete restructuring of the means by which the utility is held accountable to ratepayers.

The reluctance of city councilors to approve electricity rate increases (despite the fact that LADWP's rates are some of the lowest in the state) stems from frustration with the utility's lack of transparency and distrust among the city's residents of the LADWP's powerful labor union. Residents are skeptical of the perceived powers of the International Brotherhood of Electrical Workers Local 18, which represents LADWP workers and has long been helmed by Brian D'Arcy. Many Angelenos perceive the LADWP as an impenetrable, opaque institution completely captured by its labor union. The labor union has been an effective negotiating voice for LADWP employees but is held in suspicion by Los Angeles residents because of its ability to contribute to city election campaigns, internal union contract negotiations within the LADWP, and what are seen as inflated salaries.[35]

Tensions surfaced in 2007 when the question of rate increases was raised in city council. The LADWP sought a 9 percent increase in electricity rates, phased in over three years, to cover needed infrastructure repairs and begin funding new energy conservation and solar energy programs. The council was reluctant to approve a rate increase. Many were not convinced that the rate increases were legitimately needed for infrastructure or programs but rather to further increase the salaries of utility employees. In September 2007 the *Los Angeles Daily News* ran an article criticizing the rate increase request on the basis of the salaries of DWP employees, going so far as to post the salaries of all 8,500 DWP employees on its website, claiming they make 20 percent more than the average city worker (Barrett 2007). Councilor Greg Smith criticized the internal negotiation of union contracts at the LADWP, suggesting that the city council should take over future labor negotiations.

After making its proposal public, the LADWP held a series of twenty-five public meetings throughout the city over the course of four months. The proposal then went to the city council, framing the increase as a way to protect customers from volatility in the energy market, promote reliability, and advance environmental objectives. It included a proposal for two distinct rate zones and a tiered rate structure. The tiers encourage conservation by charging customers according to the amount of energy they use, penalizing high-energy users. The zones allowed people who live in the San Fernando Valley to have a larger base energy use allocation because of higher temperatures there.

The proposal was met with significant resistance from the city council, with concerns over how the money would be spent and what the impact of the rate hike would be on low-income residents (California Current 2007). Central city residents deemed the two zones unfair. The city council began to discuss mechanisms for greater oversight of the utility, such as a citizens committee, to be sure the extra money was spent in the manner laid out. The issue was given back to the LADWP and was not on the city council's agenda again until March 2008. It passed 11–2, with then-councilor Eric Garcetti voting no.

Villaraigosa was reelected in 2009, and in 2010 the LADWP was in need of a new rate increase. In November 2008 the city had upped its goals for solar power with the introduction of *SolarLA*, aiming to increase solar from zero to 20 percent of the city's energy supply by 2017. The LADWP estimated that to meet its renewable energy goals it needed to increase electricity rates by 5 percent to 8 percent per year through 2020, for a total of $250 million. This time Villaraigosa actively campaigned for the rate increase, making the case publicly for an 8 percent renewable energy surcharge that would be dedicated to renewable energy projects.[36]

In 2010 California's economy had been hit hard by the recession—unemployment was high in Los Angeles and wages were flat. Villaraigosa billed the rate increase as a way to bring jobs to the city (Steinhauer 2010). The city council initially rejected the increase, citing concerns about how the funds would be spent and the challenge of asking customers to pay more during a recession. Leading up to a second vote on April 15, Villaraigosa and the LADWP threatened the city council with bankruptcy, claiming the LADWP would not be able to transfer its usual portion of collected fees to the council's general revenue fund (Zahniser and Willon 2010). The council was furious over this threat, but it ultimately passed a 4.5 percent increase with an 8–5 vote.

The battle over rate increases for renewable energy and energy efficiency projects left its mark on city politics. Villaraigosa's state of the city address on April 20, 2010, referred centrally to these tensions:

> Now, I don't have to tell you that over the last several weeks we have allowed darkness to cloud our optimism. I think that you could even say that we have allowed the strain of the challenges we face to undermine civic unity. And I don't have to tell you that the past several weeks have left many people disappointed. The sparring over the budget and raising rates at the DWP became far too negative. I would be the first to say in our effort to lead a green economy in Los Angeles, I have stepped on more than a few rakes.

The battle prompted city council to put a charter amendment to a popular vote, which would establish an Office of Public Accountability and Ratepayer Advocate in the LADWP to "provide independent analysis and assessment of Department actions with respect to water and electricity rates." It passed with 78 percent of the popular vote.

While instituting the Ratepayer Advocate has served to ease tensions between the public, city council, and the LADWP, it is not clear that it has served as a panacea. Some groups, including the *Los Angeles Times*, have been highly critical, saying he has not gone far enough to protect the interest of the public. A consultant's report commissioned by the city controller in 2015 found that the Office of Public Accountability "is stuck in a 'no man's land' as it is neither a regulator nor a truly independent advisor and is still searching for relevancy," and recommended strengthening the position's political independence (Dolan 2016).

There is evidence of ongoing public mistrust of the LADWP. While rate increases have been easier to come by, there are still concerns among the public about corruption and transparency at the utility. In 2016, there was a move by the city council to increase the autonomy of the LADWP in an effort to help streamline the implementation of new programs. Both the LADWP and the Ratepayer Advocate felt that the utility's biggest problem was the complexity of the agency's bureaucracy and politics that prevented it from spending the money it was already collecting (Los Angeles Times Editorial Board 2016). As a public utility, reforming the LADWP would require a public referendum, and reform was put to a vote on November 8, 2016, in the form of Measure RRR. The measure would give greater autonomy to the LADWP and restructure its relationship to the city council. It would have expanded and professionalized the board and doubled the Ratepayer Advocate's budget. Perhaps most importantly, the LADWP would be required to prepare four-year strategic plans to be approved by the city council and mayor, and after that set rates through the board and approve programs consistent with the strategic plan without having to go through the city council. Finally, it sought to allow the city council to approve an alternative civil service standard for LADWP employees through collective bargaining and give salary-setting authority for LADWP employees to the board.

The measure was supported by the Ratepayer Advocate and the LADWP, neighborhood councils, and some environmental and community development NGOs. It was opposed by environmental and consumer rights groups and homeowners associations as well as the Fix L.A. Coalition, composed of public service unions, community groups, and social justice organizations as a power grab by the LADWP. Ultimately, the measure was defeated by 1.7 percentage points (50.87 percent to 49.13 percent), a matter of just 19,014 votes.

Los Angeles has also had to build voter coalitions to support its efforts to fund new transportation projects with the adoption of Measure R in 2008. Villaraigosa spearheaded fundraising for the "Yes on Measure R" campaign, led by the regional transportation agency, actively working to gain public support for the initiative. An even broader coalition, "Move LA," was also highly influential. The Conservation Law Foundation writes that putting together the coalition of "environmental, labor and business communities, who had never worked together before and had literally never visited each others' offices, and getting them to agree to collaborate, sent a powerful signal to decision makers . . . and ultimately provided the necessary momentum for Measure R" (Conservation Law Foundation 2010).

Coalition members included real estate developers, the Los Angeles Chamber of Commerce, and the Los Angeles County Federation of Labor. Persuading Angelenos to pay for transit has historically been very challenging, and the ballot measure received nearly the exact number of supporting votes needed, passing with a 0.4 percent margin.

Toronto's most recent efforts to govern climate change mitigation have placed mobilizing voter support at the center. Supporters within city government have recently insisted on launching a public engagement process as a way of bolstering agencies' budget requests to the city council.[37] The city's climate change mitigation efforts are being framed as a way of building a low-carbon future that includes a healthy city, good jobs, and equitable communities. This strategy arose in response to Mayor Tory's prioritization of housing and transportation but also from the city agencies' own research and discussions with other cities about best practices and effective strategies.[38] The research showed that a majority of people know about climate change and are concerned, but a much smaller group of people are willing to support and take action on climate change.

The Environment and Energy Division (EED) and TAF are working together to generate greater buy-in from the community, holding a series of public consultations and planning for more. Just over a thousand have participated in these early consultations. The hope is that demonstrating strong public support, and acknowledging the city's other priorities, might help the *TransformTO* effort to move beyond piecemeal funding allocations from city council by making a compelling case for the new initiatives. "We need a political constituency to support work like *TransformTO*, which means it has to be relevant to a broader set of people. . . . So, we need to talk about jobs, lower bills, quality of life, a better city for your children, to get a larger group to the table. We need to talk about things of interest to them."[39] *TransformTO* is explicitly aimed at mobilizing external resources and facilitating learning across divisions and between the city and the community.

New York City has learned firsthand the pitfalls of not effectively building voter support with the failure of its congestion pricing initiative in 2008. From the beginning, a number of groups expressed concern about the proposal, including that it targeted poor communities that commute by car to Manhattan and particularly those who lived outside Manhattan (S. Schwartz et al. 2009; American Road & Transportation Builders Association 2013). As was the case with the LADWP, many residents are quite distrustful of the Metropolitan Transportation Authority (MTA). Further, they "understand that the mayor of the city has little to do with the MTA, and when Bloomberg promised improved mass transit during congestion pricing, they did not believe him" (Chronopoulos 2012, 199). A poll conducted by Quinnipiac University in 2006 found that 62 percent of New Yorkers opposed congestion pricing while just 31 percent supported it (American Road & Transportation Builders Association 2013). In August 2007, support in Manhattan was 54 percent in favor, 36 percent opposed; by November support was 46 percent and opposition was 47 percent (Chan 2007). The city had made a decision not to engage many of these communities in the conversation about congestion pricing, and in the end this was the policy's downfall.[40] Rather than now having these conversations the city seems to have shifted its efforts to other means of governing transportation. De Blasio has promised a more transparent approach, but time whether and how this manifests is yet to be seen.

Capacity Building

Developing and implementing new policies and programs to reduce GHG emissions is likely to be a new task, both for city governments and for those who are targeted by their initiatives. There is likely to be substantive uncertainty about how to best design and administer new programs and how different mitigation actions shape emission outcomes. Stakeholders may lack experience and capacity in making the changes necessary to reduce GHG emissions. Building the capacities of important stakeholder groups as well as of the city's own administration and knowledge base can therefore be central to mobilizing the resources and participants necessary to change behaviors and practices. Improving administrative practices or the technical know-how of building owners, for example, empowers people and allows resources to be used effectively.

The experiences of New York City, Los Angeles, and Toronto reveal that cities are grappling with these uncertainties and working to build capacities both internally and among stakeholders. These three cities have found, often through trial and error, that it is not enough to require action, or even to change incentives. Governing climate change mitigation also requires an investment in capacity building, data collection, and education so that stakeholders, citizens, and administrators are able to meet new requirements and adopt new standards.

Government Capacities

Each city has sought to increase city government capacities in order to more effectively mobilize resources and actors for climate change mitigation, and this work has taken different forms. New York City has focused on increasing the city government's knowledge base as a means of designing more effective policies for reducing building energy use. The city realized that while energy use in buildings was clearly responsible for the vast majority of GHG emissions, they lacked a comprehensive understanding of how energy was being used in and among the city's buildings, and therefore which measures were likely to have the largest effect on GHG emissions. As one OLTPS manager told his staff, "we are now entering fact free zone."[41]

The city was subsequently the first to develop an energy use benchmarking ordinance as a way of providing baseline data for energy conservation policies and programs. The city wanted to make energy use visible because, as Bloomberg is famous for saying, "you can't manage what you can't measure." The city has sought external support for its efforts to enhance its data capacities, working with IBM on technical initiatives and receiving funding from the federal Department of Homeland Security.[42]

The city included benchmarking legislation, Local Law 84, in its 2009 package of legislation for the Greener, Greater Buildings Program. Local Law 84 requires owners of buildings greater than 50,000 square feet submit to the city annual benchmarking data for energy and water use. This law covers around 16,000 buildings in the city, which account for 45 percent of the city's total GHG emissions and more than half of the city's square footage (Mattern 2013). Building owners can assess the efficiency of their buildings, and it provides prospective buyers and tenants with more information about their purchase. In 2017 the city council passed, and Mayor de Blasio signed, legislation that expanded the benchmarking rules to include buildings between 25,000 and 50,000 square feet, which in part shifts the focus from developers to nonprofits.[43]

The second piece of capacity-building legislation was an auditing requirement, or Local Law 87. This law requires that buildings over 50,000 square feet undergo a thorough energy audit and retro-commissioning every ten years. Under this law building owners provide an assessment of the energy efficiency of their building and of the performance of energy intensive equipment and systems (e.g., heating and cooling systems). Building owners report their energy use with the U.S. Environmental Protection Agency's free software called

Portfolio Manager[44] and can also compare themselves to similar buildings in the United States. The city is working to develop a tool for easily comparing buildings within New York City as well. The hope is that by requiring building owners to assess their building's performance on a regular basis, they will become aware of opportunities for efficiency and cost savings while increasing the city's own knowledge base.

Similar benchmarking legislation has now been adopted in Los Angeles. Villaraigosa focused primarily on working through the LADWP to incentivize energy conservation and efficiency measures. Villaraigosa had a relatively good relationship with the utility, having run for mayor with the support of the union. Garcetti, on the other hand, has been a vocal critic of the LADWP and ran without the union's support. Rather than working exclusively through the LADWP, or relying on executive orders, Garcetti sought to replicate New York City's benchmarking and auditing legislation in Los Angeles.

The state of California already had a program requiring that energy consumption data be shared with purchasers of nonresidential buildings at the time of sale (AB 1103), and compliance was very low. Commercial building owners were highly resistant, and there was a reluctance to disrupt real estate deals in a city with less than 1 percent vacancy rate. While Los Angeles County reported 20,000 such transactions in 2014, the LADWP only received 20 requests for energy consumption data.[45]

Working closely with the city council and commercial sector stakeholders, the Garcetti administration slowly built support for the idea of benchmarking. In December 2014 the city council unanimously passed a motion calling for the Department of Building and Safety to convene a stakeholder process to develop legislation able to promote energy and water efficiency in buildings. From January to July 2015 the city held a series of meetings and workshops with stakeholders, and by December 2016 the city unanimously adopted the Existing Building Energy and Water Efficiency Ordinance. The new ordinance requires that buildings over 20,000 square feet register their energy and water use with the city annually. Every five years building owners must either demonstrate that their buildings meet energy and water efficiency standards or they are actively pursuing a strategy for meeting such standards. The city is currently in the process of rolling out the new program and helping stakeholders understand the reporting requirements.

Toronto also saw value in adopting an energy use benchmarking program, as it was having trouble accessing building level energy use data through their utilities. TAF learned of the benchmarking and reporting bylaw in New York City and brought in an expert from the Institute for Market Transformation in Washington, DC, to talk to them about benchmarking policies being used by U.S. cities.[46] The city convened a stakeholder group and were a year into the research process when the provincial Minister of Energy decided instead to adopt a province-wide benchmarking program. The city and province then collaborated on finishing the research and developing recommendations. In 2017 the province passed the Reporting on Energy and Water Use Regulation, which requires that owners of buildings larger than 50,000 square feet report annual resource use. The rationale is that with public disclosure of energy use the city and province will have better data for energy efficiency programs and eventually be able to develop more targeted regulations.

Looking beyond data gaps, Los Angeles has had to confront administrative procedures that prevent programs from developing as quickly or as efficiently as many would like. By 2011, it became clear that the demand for solar energy was greatly outpacing the LADWP's capacity to support installation, provide rebates in a timely manner, and guarantee customer safety. In April the utility had confirmed requests for $112 million in solar rebates but had only $30 million budgeted annually for the program. The LADWP chose to suspend the Solar Incentive Program (SIP) for five months in order to review and revise the program. They realigned incentives with state levels and improved the administrative processes in place for reviewing and approving new projects. When the SIP was re-launched in September 2011 it had twice as much funding for the following three years (increasing from $30 million to $60 million), an accelerated inspection procedure, and an online, automated application tool designed to streamline the application process.

Within five days of the re-launch, LADWP received applications for over $7.4 million in rebates to build 3.25 megawatts of solar power on Los Angeles rooftops. The utility still sees room for improving administrative procedures,[47] and the *pLAn* includes a requirement that LADWP shorten the current eight-week approval process for new projects.[48] The LADWP is still not on track to spend all of the money by the end of the program, but it plans to keep it open to help satisfy the tremendous demand for solar energy in the city.

Introducing solar energy in Los Angeles has revealed trouble in the city's building code that many cities also grapple with, including New York City. The feed-in tariff program is designed to encourage solar energy developers to build solar projects and sell the energy produced to the LADWP. Developers assumed such projects would fall under the protection of California's Solar Bill of Rights, passed in 1978.[49] Some bought residential land in the city and tore down houses to build solar energy projects. Though residents complained, the LADWP is not able to pick and choose solar energy purchases based on the location of the project. Instead, the city's Planning Department took up the issue and interpreted the zoning code as not allowing such projects and requiring a conditional use permit. These typically take between six and twelve months to acquire, which deterred developers. LADWP and private solar energy developers worked with the mayor, city council, and Planning Department to get a Master Conditional Use Permit. This qualifies projects not in agricultural or residential areas, and only requires they obtain a building permit from the city. Planning and launching new solar programs in Los Angeles has required reinterpretations of the city's building code and creative solutions to allow projects to move forward.

Energy efficiency programs in Los Angeles have also struggled to work through the influx of resources available (around $150 million to $200 million per year), revealing further capacity gaps at the LADWP. The challenge for the utility has been developing the capacities it needs to administer a budget of this size and navigating the "rigorous internal processes around spending and contracting."[50]

A manager at the LADWP sees three such challenges. One is the civil service process for hiring new staff. The LADWP can only hire if there is a permanent need for a new position, not to meet the needs of a five-year program, and the process of generating a list of verified candidates takes two years. Second is the difficulty in setting up contracts, as the union is hesitant to see things outsourced. The purchasing process requires a competitive RFP for technical services, which takes eighteen to twenty-four months to complete. Third is the utility's outdated data systems, which do not interface with one another or simply do not exist. Needed systems are often built from scratch and, given the challenges of setting up external contracts, staff are often building ad hoc data systems. As this manager put it, "when you add $100 million to a system not built to absorb it, it takes a long time."[51]

Toronto has also had to confront capacity gaps as it works to govern climate change mitigation. For example, developing its carbon credits policy requires expertise in verification, tracking, and marketing that the city does not have and has therefore contracted out to third parties. The city tracks its emission reductions on specific projects but there are steps and documentations that a consultant has to help prepare in order to produce verifiable carbon credits. The desire to maintain carbon credits has served to increase data collection and monitoring within the city. For example, agencies are eager to make sure they know the condition of a building before and after a retrofit so they can use any GHG emission reductions to create a carbon credit. "That way credits are real, measurable, and the city owns them."[52]

The city also recently hired a sustainable finance manager position in the Energy and Environment Division in order to capitalize on green banks and provincial programs and advocate for the city to these resources.[53] Through the *TransformTO* process the city has sought to raise awareness about the importance of good data and the data gaps the city faces.

According to city staff, one of the most important ways Toronto is building city government capacities is by engaging with transnational city networks such as the C40.⁵⁴ They use their contacts and networking opportunities in the C40 to identify new ideas, best practices, and financial support for their work. For Toronto, this is central to the work of city staff, who are often pursuing the city's GHG emission reduction targets in the absence of significant political or financial support from City Hall.

Stakeholder Capacities

The substantive uncertainties that come with climate change mitigation mean that requiring or incentivizing action may not always be sufficient for mobilizing participation. Stakeholders—whether they are building owners, developers, or key segments of the public—may require greater knowledge and awareness in order to comply or change their behaviors. Knowledge and capacity building can be a way of mobilizing external resources and actors as awareness and abilities grow. These three cities have recognized this as an important strategy and have sought ways to increase stakeholder capacities.

New York City has developed a strategy of pairing mandates with assistance, based on their experience with implementing new energy conservation programs. The city supplements regulatory requirements with technical support, education, and funding for implementation. When the city passed legislation for "greening the codes," they realized there was a need to support enforcement and compliance; building owners were not familiar with the new codes or technologies available for meeting them. The city saw it as its job to help solve such problems in the building industry rather than leaving it to industry to figure it out. They coordinated a series of workshops and training sessions convened by groups such as the American Energy Association; the American Society of Heating, Refrigerating, and Air-Conditioning Engineers; and the Urban Green Council using some state funding from NYSERDA. The workshops aimed to provide building owners and auditors with the information they needed to comply with the new building codes. The city established Green Light New York as a source of information resources and training for building owners and auditors, and GreeNYC as an awareness building campaign.⁵⁵ They also gave millions of dollars to the Pratt Institute to develop a "model audit" for energy efficient buildings.

The city considered the model a success and used it in subsequent initiatives. When the city council passed legislation banning the use of high-emission heating oils, they also developed a program that would help building owners transition in a way that was fast and cheap. This took the form of a series of public–private partnerships between building owners, utilities, and the city to accelerate property owners' conversions to cleaner-burning heating fuels. "The idea was to go beyond simply telling building owners what to do, to educate them about public health impacts, and in turn help them to help New Yorkers breathe cleaner air" (Charles-Guzman 2012, 46). This led to the creation of a task force that included commercial and residential property owners, real estate associations, environmentalists, technical experts, and the local natural gas utilities (Charles-Guzman 2012). The real estate industry was concerned that gas wouldn't be readily available and that conversion would be expensive and technically impractical.

The work of the task force culminated in Clean Heat, a program that has two parts. First is an awareness and education campaign aimed to convey to the public the problems associated with heavy oils and the financial benefits of switching fuel sources. The city also created its own database of the state of the boilers in ten thousand of the city's buildings.[56] Second is the creation of the Clean Heat Accelerator, which provides technical and financial assistance to building owners to assist in completing heating oil conversions and energy efficiency upgrades. In June 2012, Bloomberg committed $100 million to buildings that needed to convert to cleaner fuels. By 2015 the city had effectively phased out No. 6 heating oil, the cheapest but most noxious fuel source for the aging boilers, as a primary heating oil for New York City's buildings.

Finally, New York City coupled capacity building with its energy use benchmarking and auditing legislation. The city discovered that energy efficiency requires a workforce: people able to do the work required to implement these programs. In keeping with its view that the city should help industry find solutions to problems, it supplemented legislative actions with a significant amount of capacity building and technical support. The city sought to raise awareness among building owners of the programs and their requirements; with 16,000 buildings across five boroughs there is a significant amount of variation. The city partnered with a range of organizations—consulting firms, Urban Green Council, CUNY, ConEdison, NYSERDA, industry leaders, and professional organizations—to conduct a one-year public outreach process. This included a speaker's bureau, local workshops, and information campaigns aimed at educating building industry organizations, building owners, managers, tenants, and any group impacted by the Greener Greater Buildings Program (GGBP). These were targeted at helping people understand how to avoid a fine. The de Blasio administration has subsequently established the Retrofit Accelerator, a tax-funded, proactive one-stop-shop where building owners can get all the information and support they need for energy efficiency retrofits.

After the first year of benchmarking was completed, the city found that consultants were doing 70–80 percent of the benchmarking and of that 80 percent was being done by around thirty consulting firms.[57] This helped the city target its education efforts. The city wanted to be in conversation with this consultant community and began to put out a monthly newsletter from the mayor's office. The

OLTPS, NYC Department of Buildings, CUNY, and NYSERDA jointly launched a benchmarking help center.

The next step is working to ensure data quality. This is essential as the city works toward its goal of an 80 percent reduction by 2050, as "each sector and each building needs a goalpost. We are not in a position to mandate, but we can provide the information and targets to buildings. In the absence of mandating, that's powerful."[58] Collecting valuable and reliable data requires a well-trained work force. In collaboration with the Green Jobs Green New York program and NYSERDA, the city has developed an $8 million strategy designed to create 17,800 green construction and development jobs that can support the benchmarking and auditing requirements of the GGBP (Institute for Building Efficiency 2010).

The city funds some retrofits in private buildings, but it also tries to connect people with financial incentives from the state and ConEd. The city's "Green Building Financing," funded in part by federal stimulus money, provides owners with a direct loan, which can help mitigate the cost of retrofits and improvements required by the legislation. The loan repayments were designed to be less than the total amount of energy savings (Butler and Labhart 2010). In 2011 the city also created the New York Energy Efficiency Corporation to help businesses with energy upgrades. The corporation is designed to provide financing for energy efficiency projects and comprehensive information about funding opportunities and tax benefits, and in 2012 it provided $47 million in loans (City of New York 2013).

Los Angeles is mirroring in some ways New York City's approach as it rolls out its own benchmarking legislation. It has been holding a series of informative training webinars with building owners aimed at guiding them through the reporting and auditing process to make things easy and transparent. It has also developed a database of consultants able to be hired by building owners to complete their reports and is considering a training program through the U.S. Green Building Council.

In Toronto, TAF has again played an important role by working actively to fill key capacity gaps for stakeholders. TAF educates stakeholders about programs that are available to them and seeks to understand what their needs or obstacles are for behavior change. It also focuses on identifying gaps in skills and leadership styles necessary to lead partnerships and contribute to collaborative work.

As part of the *TransformTO* effort, the Energy and Environment Division and TAF are leading a multicriteria analysis exercise with a thirty-five-member advisory group made up of stakeholder leaders from across a spectrum of interests such as construction workers, nurses, university representatives, and interdivisional agencies. They will use the modeling exercise to develop a baseline emissions scenario and identify the gap between that and the city's goal of reducing GHG emissions 80 percent by 2050. A primary goal of the modeling exercise is to educate people and demonstrate that there are different pathways to this goal that maximize different benefits. The group will ultimately produce a recommendation document for the city council.

The Challenges and Varieties of Mobilization

Governing climate change mitigation requires city governments reduce the uncertainties inherent in reducing urban GHG emissions. This work underpins, but also extends beyond, the task of passing new bylaws or introducing new incentives. New institutional, political, and administrative arrangements are needed for resources and actors to be mobilized in pursuit of the changes necessary for reducing emissions in the city.

New York City, Los Angeles, and Toronto have recognized the importance of the governing strategies of institution building, coalition building, and capacity building and have pursued them in a variety of ways (table 4.1). The cities have used institution building to facilitate coordination among disparate actors and ensure the longevity of climate change mitigation programs. Mayoral powers and city council support are important for how city governments build such institutions. In New York City's case, the powerful mayor's office allowed Bloomberg to build an office capable of coordinating climate change actions throughout city government. Support from city council ensured that the aims and structures became hard-wired into the city.

Los Angeles and Toronto sought out different means of institution building, as the mayors did not enjoy as much control over the civil service or support from city council. In the case of Los Angeles, both Villaraigosa and Garcetti have worked to create a network of leaders in the civil service able and willing to embrace climate change mitigation goals. Villaraigosa placed coordination responsibility in the city's Environmental Affairs Department, which proved vulnerable during the economic downturn. Garcetti is coordinating the work through his Office of Sustainability, but the group is not particularly well-resourced. Toronto placed climate change mitigation within the Energy and Environment Division, a group without authority or mandate to coordinate the work of other agencies. Rather, the city's dedicated climate change organization, TAF, has played a valuable role in coordinating the activities of various groups. Codification has helped ensure climate change mitigation work continues in Toronto, despite large swings in mayoral commitment to the issue.

Coalition building has also been instrumental to activating the governing abilities of these city governments, highlighting the political work required to move the climate change agenda forward. New York City, Los Angeles, and Toronto have been adept at building and leveraging coalitions of influential stakeholders supportive of the city's work on climate change. These coalitions are fairly diverse, going beyond core environmental constituencies. Labor and business groups, such as unions and chambers of commerce, are often supportive of the job- and investment-creating potential of local energy efficiency and alternative energy generation projects. Building owners and real estate organizations are beginning to see the advantages of energy efficiency for reducing costs and increasing the value of commercial real estate space.

Such coalition building has varied in its formality and consistency. In New York City the use of stakeholder task forces and advisory groups has provided access to decision making and helped to create longer-term relationships among participants. Coalition building in Los Angeles and Toronto has been more ad hoc, as support and resources have been needed for particular initiatives. Toronto's new *TransformTO* initiative is seeking to build a more embedded and stable coalition of key stakeholders.

Such coalition building is important at multiple stages in the policy process, from developing policy proposals themselves (as in the case of New York City) to helping policies to pass through the city council (as in the case of Toronto) or by popular vote (as in the case of Los Angeles). Coalitions for climate change mitigation are diverse, can be more or less stable, and can have a broader or narrower focus.

Building and mobilizing voter support is a means of coalition building that the three cities recognize as essential to governing climate change mitigation but have made different levels of progress in achieving. Los Angeles has been able to mobilize voter support for new funding for climate change programs, specifically raising electricity rates and raising the sales tax. The importance of this strategy for the city comes in part from the fact that the city is serviced by a municipal utility and therefore electricity rates are heavily shaped by public debate and opinion. This support was hard won and has required significant changes to the institutions and procedures involved in delivering electricity services in Los Angeles. The LADWP has now instituted a Ratepayer Advocate as a means of increasing transparency and accountability. There is also greater emphasis placed on the importance of a public dialogue around rate increases and the ways that additional revenue can best be used. Public understanding and support (or, in many cases, skepticism) is a very tangible constraint on what the LADWP is able to accomplish and, therefore, the progress Los Angeles is able to make on its climate change mitigation goals.

In Toronto, a vocal and supportive public has been a key source of political mobilization. Building this support is also an explicit goal of the new *TransformTO* plan. New York City continues to make the case to the public for leadership and investment in climate change mitigation; the recent approval of congestion pricing could mark a shift in public sentiment.

Capacity building has helped the cities address gaps in knowledge and administrative capabilities both in the city and for stakeholder groups central to implementation. New York City has made this an explicit priority, investing in data collection programs, and training and education initiatives for building owners and auditors. Capacity building has become part of the city's standard operating procedure, as they pair mandates with assistance in many of their efforts.

Los Angeles suspended implementation of its Solar Incentive Program in order to work through bottlenecks in administering the funds. They are also beginning to take more seriously the need for stakeholder education and training for energy efficiency programs, providing a webinar series for the benchmarking ordinance. Toronto staff are keenly aware of the need for capacity building to fuel their work. One strategy they are using to target stakeholders is the use of stakeholder-driven scenario development, which provides stakeholders with an in-depth understanding of the mechanisms that drive GHG emissions, options and tradeoffs for their reduction, and the importance of good data. Staff have begun to rely on intercity networks such as the C40 for best practices, new knowledge, and opportunities to finance city programs. In both Los Angeles and Toronto capacity building has been driven by the civil service rather than elected decision makers.

Developing these governing strategies has been an evolving process in the three cities, in some cases learning through trial and error. For example, in Los Angeles political resistance to various initiatives at the LADWP helped to reveal the issues the utility faced with regard to transparency and credibility and forced decision makers to reform decision making to gain the public support needed for important initiatives. Introducing the Ratepayer Advocate had not been part of the city's original strategy for increasing the use of solar energy, but it became critical to effectively governing as the conversation unfolded and became more public. In Toronto, the need to mobilize a broader group of stakeholders for climate change mitigation programs has intensified as mayoral support for these initiatives has dwindled. What once might have been justifiable based on environmental outcomes alone now must be clearly tied to economic and employment benefits. Capacity gaps are often only apparent once a new program or policy is on the table, highlighting a lack of knowledge or awareness on the part of the city or stakeholders. Climate change mitigation governing strategies are therefore dynamic but consistently target the institutions, coalitions, and capacities that underpin the city's work to reduce GHG emissions.

TABLE 4.1 Examples of the use of three steering strategies in New York City, Los Angeles, and Toronto for climate change mitigation

STEERING STRATEGY	PURPOSE OR TARGET	NEW YORK CITY EXAMPLE	LOS ANGELES EXAMPLE	TORONTO EXAMPLE
Institution Building	Coordination	Well-staffed Office of Long Term Planning and Sustainability	Inter-utility agreement between the LADWP and SCG	Toronto Atmospheric Fund
	Longevity	Codification of targets, planning, and reporting requirements	Mandates and targets for the LADWP	Council approval of targets and measures to be taken
Coalition Building	Stakeholders	Use of Green Codes Task Force to develop recommendations	Prioritization of local solar projects to generate local benefits	Framing energy efficiency as revenue neutral, source of job
	Voters	*Identified but poorly developed*	Reform of the LADWP to support rate increases; Leadership on passing Measure R	*Identified but poorly developed*
Capacity Building	City Government	Benchmarking legislation for data collection	Revising administrative systems for solar energy programs	Tapping intercity networks like the C40
	Stakeholders	Training and education programs for building owners and auditors	Webinars for building owners on new energy use reporting requirements	*TransformTO*'s stakeholder advisory group work

New York City, Los Angles, and Toronto have each developed their own strategies for building institutions, coalitions, and capacities. Despite differences in political leadership, existing capacities, or broader policy context, each city has sought ways to build the institutions, coalitions, and new capacities able to support their work on climate change mitigation. In each case there was a need to restructure these arrangements to accommodate and incorporate the goal of reducing GHG emissions, and the cities have each recognized the importance of this restructuring. Working within their unique contexts, they have used institution building, coalition building, and capacity building as strategies for mobilizing the resources and actors needed for climate change mitigation. This underscores Castán-Broto's argument that urban climate change mitigation efforts can reveal not just the intentions or aspirations of city governments but also the "underlying paradigms and the contexts of intervention in which governing actors operate" (2017, 8).

Political commitment from City Hall is instrumental in shaping the governing strategies cities use to mobilize actors and resources toward reducing GHG emissions. Such commitment has been strongest, most consistently, in New York City and this is evident in the city's investment in climate change governance. New York City has taken the most significant steps toward hard-wiring climate change mitigation into city institutions. The city has also invested considerably in stakeholder participation, input, and capacity building. Funding for such efforts has come from the city's own budget in many cases, though at times it has sought external funding for its work. This type of prioritization and investment has come from the buy-in and commitment of not just the mayor but city council—and a willingness of the two to work together for the betterment of the city.

Los Angeles and Toronto have struggled to maintain consistent political leadership on climate change. While both Villaraigosa and Garcetti have made climate change mitigation a priority for their administrations in Los Angeles, city council has often been a reluctant partner at best. This is due in part to the centrality of the LADWP for reducing emissions in Los Angeles, an entity that has long been a political flash point for the city.

Tensions around climate change mitigation in Los Angeles have also been the product of more basic political disagreements and rivalries. While in city council, Garcetti was often a roadblock for Villaraigosa's environmental initiatives, leading the charge to eliminate the Environmental Affairs Department. Given his current commitment to climate change mitigation, this could be read as more political than ideological. Ultimately, the need to resolve tensions around the LADWP will likely lead to further needed reforms of the utility, and there are now signs of greater cooperation between the city council and the mayor on climate change mitigation as the city's profile in this area grows.

In Toronto, the lack of political leadership on climate change mitigation has resulted in the "outsourcing" of many governance strategies. TAF, an organization largely external to the city's administrative arm, has taken on many governance tasks. It actively seeks out ways of facilitating coordination and filling capacity gaps both within the city and for stakeholders. City staff in Toronto rely more heavily on the input and resources of the C40 for their own work due to a lack of investment in internal capacity building. Many of the city's most promising efforts to develop its governing strategies more fully are tied to the *TransformTO* process, which is currently receiving piecemeal funding from the city council. While Toronto's creativity and unique approach to leveraging the three governing strategies speaks to its flexibility and committed bureaucrats, it also highlights the lack of political leadership on climate change in the city.

Still to be determined is what these efforts to govern climate change mitigation have accomplished. The cities have developed new and unique policy agendas and are investing in building the institutions, coalitions, and capacities to mobilize the resources and participants needed for them to launch. But to what end? Are these cities now moving in a new direction? Are they on track to realize the kinds of transformations required to meet their mitigation targets? I take up these questions in the next chapter by examining each city's progress on reducing GHG emissions, the broader changes that are attributable to their work on climate change mitigation, and the new directions each city is currently pursuing.

5

ARE WE THERE YET?

Identifying and Evaluating Urban Progress on Climate Change Mitigation

New York City, Los Angeles, and Toronto have each developed unique policy agendas for climate change mitigation. They are using the governing strategies of institution building, coalition building, and capacity building to mobilize the resources and actors needed to pursue these agendas. The cities are navigating the dynamic complexities and uncertainties of reducing GHG emissions. Their efforts reflect the underlying principles of urban governance—leveraging formal and informal authorities, acting both independently and in partnership, in order to take their cities in a new direction.

In this chapter I evaluate the progress the cities have made. Emissions in the three cities have declined between 12 percent and 26 percent since 2007. Given the uncertainties in calculating urban GHG emissions and their change over time, and the complex set of factors that shape urban GHG emissions levels, these are not qualitatively distinct levels of reduction. These figures show that the cities are not dramatically over- or under-performing but are successfully taking the first steps toward their larger and longer-term climate change mitigation goals.

As expected, linking the actions of the city governments directly to these changes in urban GHG emissions is challenging, particularly for cities that keep sporadic or inconsistent records. There are also factors beyond climate change policies and programs influencing annual GHG emissions—including population growth and weather patterns—that can mute the effects of the cities' efforts. Accounting for the effects of local climate change action must also go beyond that city's GHG emissions. Each city's governing efforts have produced broader system changes—or catalytic effects—by providing lessons or motivation for other cities or levels of government to act on climate change.

In this chapter I also examine where the three cities are aiming to go next with climate change mitigation and the adjustments to their governing strategies that will be necessary to get them there. The cities find themselves at a crossroad: they have been successful in staying on track to meet their short- and medium-term GHG emission reduction goals, but they face heightened uncertainty and greater political and institutional challenges to more significant reductions. I discuss ways the cities can govern toward more fundamental transformation.

Reductions in Urban GHG Emissions

Despite early and consistent action on climate change mitigation by these three cities, limited data and reporting make tracking changes in their GHG emissions over time challenging. New York City has done an excellent job of monitoring annual GHG emissions, but this has not been the case in Los Angeles or Toronto. It is also difficult to directly compare reductions between the cities, as they each use a different baseline year and, often, inventory methodologies. Nonetheless, current estimates reliably present the broad contours (if not precise details) of the changes in GHG emissions in each city over the last ten years and the primary contributors to those changes (table 5.1).

The trends in GHG emissions in New York City, Los Angeles, and Toronto reveal that urban climate change governance is working at a pace and scale that can be interpreted as proof of principle. The extent to which city government intervention is the direct cause of GHG emission reductions varies between cities, but in each case the city's commitment to climate change mitigation played a role in facilitating change. In each case the cities have had more success in reducing GHG emissions from city government operations, further demonstrating the effectiveness of their policy agendas and governing strategies. Governing to reduce GHG emissions in cities has very tangible results and at a pace that is in line with the goals the cities initially set out.

While the GHG emission reductions in the three cities are the product of unique mixes of policies and programs and changes in different sectors, the two primary contributors to direct reductions have been *removing coal from the energy supply* and *capturing methane emissions*. These are large, one-off projects that achieve big reductions in a short period of time. They are supply-side solutions, which have been a priority for New York City and Toronto. It is likely that demand-side interventions (e.g., energy efficiency and conservation) will take longer to produce tangible reductions in GHG emissions. Changes to transportation networks and mobility patterns are not yet major contributors to urban GHG emission reductions. Again, investments in new infrastructure will take time to manifest, such as the projects slated for completion through Los Angeles's Measure R. Reducing GHG emissions from transportation must be a priority for the cities going forward.

New York City

Tracking New York City's progress in reducing GHG emissions is a relatively straightforward task. New York City stands out from nearly every other government (city or otherwise) in its systematic and transparent inventorying of GHG emission sources and trends.[1] This is due in no small part to Bloomberg's ubiquitous motto, "you can't manage what you can't measure." Using peak GHG emissions from 2005 as its reference point, the city's goal was to reduce GHG emissions 30 percent by 2030. In 2012 the city's emissions had fallen by 19 percent, their lowest levels since 2005. Emissions crept up slowly in 2013 and 2014, due to cold winters and population growth, such that the city's most recent GHG emission inventory estimates an 11.7 percent net reduction below 2005 levels. The city has reduced emissions by nearly 10 megatons, but population growth and cold winters have partially blunted the net effect of these reductions. Per capita emissions have declined by 17 percent, from 7 tons to 5.8 tons, as the city has gained population (4 percent) and increased economic production (15 percent) during this time.

Most of the city's emission reductions (nearly 70 percent) were achieved prior to 2012 as a result of power plants within and outside the city switching from coal to natural gas. Natural gas now accounts for 34 percent of the city's GHG emissions, due to both the switch in fuel source and increasing conservation of electricity. Indeed, electricity conservation accounts for around 10 percent of the city's GHG emission reductions. Fewer cars on the road, reduced waste generation, and landfill methane capture account for the majority of the remaining emission reductions.

City government emission reductions followed a somewhat different pattern. While overall the rate of savings was similar (13 percent), electricity efficiency played a much greater role, accounting for 16 percent of reductions (compared to 10 percent citywide). The two biggest drivers of city government emission reductions were efficiency in electricity generation (31 percent) and capturing methane at landfills (16 percent). Prior to 2014, cleaner electricity supplies drove reductions, but recently declining use of electricity and heating fuels have been driving changes in emissions. The use of No. 6 heating oil decreased 64 percent over this time, and gasoline use in city vehicles decreased 23 percent.

Switching to natural gas produced meaningful reductions to New York City's citywide GHG emissions early on. The city is subsequently on track to meeting its goal of reducing emissions 30 percent by 2030. The city contributed indirectly to this switch by working actively to promote its interest in a lower emissions energy network. It has worked in partnership with natural gas utilities to capture methane emissions; with regulators, utilities, and stakeholders to accelerate the development of natural gas transmission lines and distribution upgrades in the city; and with the state to ensure that generation efficiency is encouraged. The city's enabling efforts have helped to facilitate the switch to natural gas by private utilities and customers.

While the city played a role in facilitating the switch to natural gas, the majority of its governing efforts have focused on increasing energy efficiency in buildings and these have not yet produced proportional reductions in GHG emissions. Reducing energy use in buildings is an inherently long-term and tedious process. Whether through benchmarking energy use or changing the building code, multiple sets of actors must shift their behavior and will encounter opportunities to do so at different points in time. Gronewald (2010) describes the process of greening the city's codes as "a blueprint for a slow, quiet, decades-long process to meet city goals for steep cuts in energy consumption and waste by 2030." Ultimately, these changes are estimated to generate a 5 percent reduction in the city's GHG emissions by 2030, the largest contribution of any single program (Urban Green Council 2012). These code changes have also contributed to fewer flooding incidents in upgraded buildings in Queens and Brooklyn.[2]

The city's benchmarking program is also showing promise, with a 75 percent compliance rate among affected building owners, and around eighty energy audits and fifty comprehensive building retrofits each year (City of New York 2013, 2014). Much of this compliance is driven by the potential to save costs and the boost in public image that comes from being seen as a sustainable building or company.[3] "GHG reductions are icing on the cake. Savings is the driver, but . . . they announce that they've reduced GHGs. It can have a major impact, it keeps you on the list of good buildings to be in."[4] For some building owners, public benchmarking was an opportunity to highlight work they were already doing and they only needed to submit a report. For others, it was the spark they needed to begin efficiency retrofits.

The U.S. Department of Energy recently completed a report that confirms the early successes of New York City's energy use benchmarking policy (U.S. Department of Energy 2015). They find the real estate community in New York City is aware of the policy and expects it to become increasingly important for the market. They also find that the building owners reporting through the program have reduced their own GHG emissions by around 10 percent between 2010 and 2013, and the program has created more than 7,000 jobs. Other estimates show the city's larger suite of programs encompassed by the Greener, Greater Buildings Programs have saved New Yorkers $700 million in annual energy costs and created 18,000 jobs (Center for Clean Air Policy 2009; Institute for Building Efficiency 2010). Some in the city see policy improvements as the primary benefit of the benchmarking program.[5] Access to better data about energy use in buildings allows the city to develop more targeted conservation policies.

One unexpected outcome of the city's governing efforts has been the conversion of the taxi fleet. The city had tried to impose requirements on taxi companies to switch to hybrid vehicles, a move ultimately deemed by a state court to be outside the city's jurisdiction. The requirements were in place long enough, and the cost-savings argument was convincing enough for some, that the city has nevertheless seen a real shift in the composition of its taxi fleet. New York City also developed a suite of financial incentives for transitioning. As early as 2009 the city found that 15 percent of the taxi fleet bore the green apple logo indicating fuel-efficient vehicles (from 500 to 2,000), and today that number may be as high as 65 percent.[6] While new initiatives might confront unanticipated barriers, putting forth the proposal and its merits can itself change behavior by drawing attention to the costs and benefits of the problem.

There are also signs that New York City's efforts to facilitate distributed solar energy generation are working, especially since de Blasio made this a focal point for his work on climate change mitigation. The city's own solar energy generating capacity more than tripled since 2013, increasing from 25 MW to 96 MW, with more than 3,500 new solar installations primarily on residential properties (Office of the Mayor 2015b). This is roughly the equivalent generating capacity to power 96,000 homes in the city. The city also estimates the solar energy programs have created 2,700 new jobs. As Los Angeles found, the primary obstacles to greater uptake of solar energy have been administrative procedures and subsequent uncertainty for solar energy developers.[7] ConEd, like many utilities, requires customers pay a stand-by rate even if they are fully solar-powered. The city still needs a turn-key approach to solar power installation, and legislation has recently been introduced by Councilor Richards that would create an Office of Solar Energy in the city able to streamline the permitting process and provide information about the rules, fees, and necessary paperwork.[8] Capacity building such as this will help facilitate behavior change and better mobilize resources for expanding solar energy development.

Los Angeles

Tracking carefully the changes in GHG emissions in Los Angeles is a tricky proposition. The city has struggled to maintain a consistent record of citywide GHG emissions. Some departments and agencies have been reluctant to share data, and decision makers have not prioritized full accounting. "The steps taken to quantify emissions or not have gotten very political."[9] This has been due in part to the city's unique position of having an airport, energy utility, and port, all of which contribute significant GHG emissions. There is concern that comparing Los Angeles's GHG emissions to other cities is "like comparing apples and oranges."[10] Internally, city departments are reluctant to reveal themselves as large emitters. Los Angeles also lost some capacity for generating and maintaining a GHG emission inventory in 2008 when the Environmental Affairs Division was dissolved. Without formal oversight or requirements for producing a GHG inventory, it has often been left for another day.

The city's original climate change plan, *Green LA*, reported the city's GHG emissions as 50 megatons in 2004. When Villaraigosa left office in 2013 he claimed that emissions had reduced by around 28 percent.[11] The city's reported GHG emissions varied during this time. In 2010 it reported to the C40 that citywide emissions were 38.7 megatons, and that this represented an increase from 2004. In 2011 the city reported 13.39 tons of GHG emissions per capita to the World Council on City Data, equivalent to around 51 megatons in total. In 2013, Los Angeles reported citywide GHG emissions to the C40 as 33.9 megatons, a 12 percent decrease from its C40 numbers in 2010.

Like New York City, Los Angeles has now adopted the BASIC inventory protocol but uses 1990 as its baseline year, mirroring international agreements. In 2016, the city estimated citywide GHG emissions had dropped from 36.3 megatons in 1990 to 29 megatons in 2013, a 20 percent reduction.[12] Using these numbers, the city's per capita emissions have decreased 28 percent, from around 10 tons to 7 tons.

These updated figures do not specify GHG emissions from city government operations, nor are there goals for these emissions in Mayor Garcetti's climate change plan. In 2007, the city set a goal of reducing emissions from city government operations 35 percent below 1990 levels by 2030. According to documents submitted by the city to the Carbon Disclosure Project, Los Angeles reduced city government emissions from 817,988 tons in 2011 to 613,473 tons in 2015, a 25 percent reduction.

Pinpointing the sources of GHG emission reductions in Los Angeles is problematic because of the city's struggle to maintain a consistent record of citywide GHG emissions. The vast majority of the reductions reported by Los Angeles appear to be due to the changes in the city's energy supplies and improvements in energy conservation. The figures from 2016 indicate the city has reduced GHG emissions from energy by nearly 5 megatons or 21 percent; waste and transportation each account for around 1 megaton. For waste this amounts to a

68 percent reduction since 1990. These figures are consistent with the perceptions of interviewees, who expected that renewable energy deployment and energy conservation were the largest contributors, followed by improvements to city government operations (e.g., waste management).[13] They expected emissions had also decreased some during the recession, particularly as people drove less for work.

Moving away from coal was the low-hanging fruit for Los Angeles, a city with its own municipal energy utility and heavy reliance on out-of-state coal-fired power plants. The city has increased its use of renewable energy sources—wind and solar—from 3 percent in 2003 to 25 percent in 2017, meeting the goal Villaraigosa set for the city in 2008 and a statewide mandate set in 2011. The LADWP has provided nearly $300 million in rebates and supported the installation of 168 megawatts of solar power through 22,000 systems. The city's goal is to move away from coal entirely by 2026, though navigating the contractual relationships involved is proving more challenging than originally thought.

The city aims to be off of the Navajo coal-fired generating station by 2019, but it will not be able to end its contracts for coal-generated electricity with the Arizona and Utah generating stations until 2025. Ending these contracts raises a moral dilemma for the city: the city negotiated for the complete shutdown of one-third of the Navajo generators, which means a loss of jobs for people in that community. Some in city council are asking the LADWP to explore options for ending contracts with the Navajo station but still providing jobs, perhaps through solar or wind. "Each Navajo job supports ten people. We want to recognize the fact that they've been good neighbors, they've supplied power for the city for forty years."[14] The city's efforts to move away from coal reveal some of the distributive issues that arise with climate change mitigation and create additional political and economic demands on the transition process.

The city's energy efficiency investments have so far resulted in a 10 percent reduction in energy use, and the city is aiming for 15 percent by 2020. The LADWP's 2016 Integrated Power Resource Plan shows they missed their energy efficiency installation targets in 2014, 2015, and 2016, likely due to administrative hurdles and backlogs. Nonetheless, the city's energy efficiency programs have reduced GHG emissions and fostered job growth in the city. Since 2014, energy efficiency programs in Los Angeles have produced an average of 16 job-years per million dollars invested (for reference, residential construction produced 10.7 job-years per million dollars) (DeShazo, Turek, and Samulon 2014). The city estimates that green building, water infrastructure, energy efficiency, and transit programs have together produced 20,000 jobs in the city (City of Los Angeles 2015).

Toronto

Toronto regularly reports its GHG emissions but has struggled to maintain updated and accurate figures in these reports. The city's most recent inventory, released in February 2019, shows Toronto's GHG emissions have declined by 26 percent since 2007 and by 33 percent since 1990. Per capita emissions for the city have fallen more quickly, declining by 31 percent between 2007 and 2016 due to population growth and continued GHG reductions. City government emissions, representing only around 6 percent of the city's total GHG emissions, have declined at a similar rate to overall emissions, by 19 percent since 2004.

One challenge in assessing the city's progress is that Toronto did not update its transportation emissions between 2008 and 2019 due to resource constraints in the city's Transportation Services Department.[15] Recent estimates being done by the city using new data from the province and academic partners indicate that transportation emissions are declining, largely due to stricter fuel efficiency standards for vehicles.[16]

The two largest contributors to GHG emission reductions in Toronto are the removal of coal from the city's energy system and methane capture at city landfills. While Toronto uses around the same total amount of electricity in 1990 as it does today, the emissions associated with this use are 60 percent lower today as the provincial energy grid went from 25 percent coal in 2003 to a complete phaseout in 2014. Coal has been replaced by expanded nuclear generating capacity, new wind and solar energy projects, and biomass-fueled generating stations. While these changes were the direct result of provincial policy, many in Toronto believe that city programs facilitated the shift away from coal: Toronto had been reducing energy use significantly through projects such as the deep lake water cooling system[17] and various energy efficiency measures.[18]

The city has played a more direct role in reducing emissions through methane capture at city landfills. Toronto has invested in methane capture technologies in the city's five landfills, which have reduced waste-related emissions by 25 percent since 2004 (City of Toronto 2015). Methane capture at two of the city's largest landfills are responsible for just over 10 megatons of GHG emission reductions, representing 5 percent of the city's overall reductions.

Toronto is also making progress with its energy efficiency programs. The Better Building Partnership has taken 131 megawatts off of peak energy use and supported retrofits in about 2,100 buildings in the city.[19] The Home Energy Loan Program is now three years old and is being used by fifty-three households for an average savings of $500 to $800 per year. The city still struggles to make significant inroads on energy use in commercial buildings. Toronto's voluntary green building standards provide recognition and some incentives for voluntary action. Overall, commercial sector uptake of the city's various programs for energy efficiency retrofits has been slow. Potential partners are skeptical of the process and of working with City Hall.[20] University partners have often chosen to undertake individual building projects rather than invest in campus-wide retrofits.

The commercial sector is the primary target of Toronto's new carbon credit initiatives. As markets shift toward green investments and corporate social responsibility, the commercial sector may be a very important partner for the city going forward. The city is working to better demonstrate the business case—or return on investment—that can be had from subsidized energy efficiency retrofits.

Catalytic Effects of Urban Climate Change Mitigation Governance

The climate change governing efforts of New York City, Los Angeles, and Toronto have had broader consequences for the cities and the collective enterprise of climate change mitigation. From changes in governance structure to a culture of sustainability to the policy choices of other levels of government, the work these cities are doing to steer their cities in a new direction is contributing to changes beyond urban GHG emission patterns. Governing climate change mitigation has forced the cities to confront political tensions around energy and infrastructure planning; to rethink administrative structures and their relationship to the private sector; and to share and learn from other cities in new ways. As leaders on climate change mitigation, these cities are also charting a path for other levels of government to follow and build on. They are contributing to the larger project of reducing global GHG emissions by testing new ideas, demonstrating the feasibility of local action, and changing the political landscape of climate change mitigation policy.

As cities build new institutions, coalitions, and capacities for pursuing their climate change mitigation goals, broader political and institutional changes are inevitable. These have manifested in different ways in the three cities, and the variation is driven by (1) the different ways the cities have used the three governing strategies; they have built different institutions, coalitions, and capacities as a result of their work on climate change, and (2) the cities' unique positions in relation to their state and provincial governments, and within larger intercity networks. Their climate change mitigation governance therefore has different channels and pathways of influence. As examples, while New York City has contributed to a shift in the business community's perception of sustainability, Los

Angeles has institutionalized greater transparency at the LADWP; Toronto has made it easier for the province to pursue its own set of mitigation policies.

These results underscore, and provide nuance to, previous arguments about the catalytic role city governments can play in the larger enterprise of reducing global GHG emissions (Hoffmann 2011; Bulkeley et al. 2014). The true GHG emission implications of the governing efforts of these three cities extend beyond their borders. Their leadership has sparked commitment and policy change in other cities and at other levels of government. By sharing experiences through the exchange of staff members, intercity networks, and formal policy channels, each city has contributed to climate change mitigation in ways that cannot be captured in GHG emission inventories alone.

New York City

New York City's climate change mitigation governance has changed behaviors, generated buy-in, mobilized a range of resources, and positioned New York City as an environmental leader. The political commitment to climate change of New York City officials has been embedded in the city's formal institutions—legislated and codified—as well as the informal institutions surrounding city practices and identity.

The Green Codes Task Force in particular has had longer-term implications for governance norms of collaboration and problem solving. Mayor Bloomberg went on to use other task forces as he pursued his climate change agenda, such as the Climate Change Adaptation Task Force (2008) and the Building Resiliency Task Force (2013). Mayor de Blasio has also adopted the task force model, using the Green Buildings Technical Working Group to inform *One City Built to Last*. City staff members were key in convincing the de Blasio administration to seek out stakeholder support.[21] The group has a different makeup than Bloomberg's task forces, more heavily weighted toward technical expertise, and is expected to serve as a sounding board and source of feedback for the administration rather than a source of policy proposals. The new working group's ability to generate ideas and recommendations that are adopted by the city may depend on the mayor's commitment to leadership in this area, the legitimacy assigned to the collaborative process itself, and the technical competency of the recommendations. Here both form and function will play a role in continuing to strengthen the coalition behind mayoral climate change initiatives.

The city's efforts have also permeated the wider community such that sustainability and climate change mitigation are now seen by the business community to be legitimate aims. Before *PlaNYC*, sustainability wasn't seen to be appropriate for serious businesses or business people, but after Bloomberg's work on *PlaNYC* the views changed. This was due in part to Bloomberg's popularity and credibility with the business community; they emulated his priorities. "Businesses were falling over each other to do sustainability things. If you wanted Bloomberg to speak at your event, do something with sustainability and he'll come to announce it."[22] City residents and employees want the local benefits of clean air and livability, and the business community sees opportunities to market their buildings as unique and their companies as environmental leaders.[23] While not every mayor will be a business leader, a clear champion from the business sector may prove useful for encouraging buy-in and enhancing legitimacy.

Catalyzing climate change action in other cities in the U.S. and abroad has always been an explicit goal for New York City, and there is some evidence that they are making progress on this. "They did a service to the nation by showing how cities can lead. They weren't the first but they were the biggest and flashiest. Their work had teeth and was followed with legislation, a new office and government structure. They led the way and legitimized it for other cities."[24] Several of the city's programs have been replicated elsewhere, including the benchmarking legislation. In addition to Los Angeles and Toronto, cities such as Philadelphia, Boston, Chicago, Atlanta, and Austin have all adopted legislation requiring large buildings report their energy use.

New York City's approach—developing an integrated, high-profile sustainability plan and institutionalizing the targets and reporting requirements—have also been adopted by a number of U.S. cities. While New York City was not the first city to make a commitment to climate change, they were more proactive than most in reorienting city government priorities and institutions in order to govern effectively, and other cities often look to them for guidance.

The City Energy Project, launched by the Natural Resources Defense Council, has taken up the task of transferring the lessons from New York City. The project's explicit aim is to help other cities develop energy use programs for buildings that are modeled after New York City's. Bloomberg Philanthropies, the Doris Duke Foundation, and the Kresge Foundation have funded the project. The City Energy Project provided both technical and staff support to Los Angeles in 2016 as it worked to pass its benchmarking ordinance.

Michael Bloomberg gained a global profile on climate change issues as the influential chair of the C40 network from 2010 to 2013, the president of the C40 Board of Directors and Special Envoy for Cities and Climate Change to the United Nations Secretary General, and now the United States Special Envoy for Climate Action. Through these venues Bloomberg has been highly influential in infusing his style and approach to urban climate change mitigation into global protocols and strategies (Gordon 2015).

Following the release of *One City: Built to Last*, New York City undertook a series of stakeholder engagement processes to reflect on current progress and devise ways to move forward toward the goal of reducing GHG emissions 80 percent by 2050. "We left some questions unanswered about getting past the easier steps."[25] One output from this process is *New York City's Roadmap to 80×50* (City of New York 2016). The main message in this report is that the projects that significantly reduced the city's GHG emissions—repowering generating stations or capturing methane emissions—cannot be replicated, and much of the low-hanging fruit has been picked. The city needs to significantly scale up its current efforts, particularly around reducing energy use in buildings, and target more aggressively the transition to renewable energy and a low-emission transportation system.

Los Angeles

Governing climate change mitigation has also produced catalytic effects in Los Angeles. The process of raising electricity rates to fund new programs highlighted the political tensions around the LADWP and forced greater transparency onto the utility. The Ratepayer Advocate now provides an institutionalized mechanism for third-party oversight and public accountability for the utility. Developing new programs for local solar energy also forced the city to be more transparent about the process and rationale behind such programs. Public awareness—and skepticism toward the LADWP and its union—remains high in Los Angeles, and there is increasing demand for climate change mitigation initiatives that are democratic and equitable. Efforts to reform the LADWP more fully have been met with public resistance, and more work is needed to engage the public in a constructive conversation about the future of the utility.

One mechanism for such a conversation is the city's neighborhood councils, which have been some of the most vocal critics of the LADWP but also willing collaborators when the opportunity presents itself. In 2005, the LADWP and Los Angeles Neighborhood Councils signed a Memorandum of Understanding that commits the utility to providing adequate notice and engagement opportunities in advance of rate increases, budget issues, new projects, and changes to policy, and it established formal liaisons between the LADWP and neighborhood councils. This relationship is a unique opportunity for the utility to facilitate a larger conversation about not just the city's energy future but also the preferred means of achieving it.

The city's efforts to govern climate change mitigation have also facilitated political and governance shifts in the city. People in City Hall report an improving willingness among agencies to work together on climate change goals. One example was a meeting of the chief sustainability officer representatives from each agency. The Department of Street Services wanted to sample new sustainable pavement on a parking lot and the Parking Department volunteered.[26] A relatively small gesture, it was seen as a promising improvement.

While the *Green LA* coalition itself no longer exists, its work helped to solidify the importance of environmental work in city politics and policymaking. Several members of the coalition remain in city government and others have fostered the creation of other influential organizations, including Climate Resolve. "Good coalescing came out of it."[27] The *Green LA* coalition that catalyzed the city's original climate change plan helped to further embed environmental priorities in city government.

Mayor Garcetti is working actively to change the role of the mayor in charting a course for the city. He is being proactive and explicit about leveraging his formal powers to forward the agenda in the *pLAn*. Garcetti wants to change the city's weak mayor culture and "implement his policies as chief executive."[28] He is also working to raise the city's national and international climate change profile, initiating the Mayors National Climate Action Agenda together with the mayors of Philadelphia and Houston. It is likely no coincidence that one of the goals of this network is to forward standardized protocols for GHG emission inventories, one of the most challenging aspects of climate change mitigation governance for Los Angeles.

Garcetti has also introduced a greater emphasis on data and data transparency in the city. His Data LA initiative provides an online portal for access to information on a number of key social, economic, and environmental indicators including GHG emissions, employment statistics, and city budgets. "It's a nerd revolution."[29] The website that hosts the data has been slow to develop, but the number of departments focusing on the indicators in the planning and programming is increasing.[30] The initiative is a huge undertaking and reflects Garcetti's internalization of Bloomberg's emphasis on measurement; indeed, it was funded in part by Bloomberg Philanthropies.

One unanticipated consequence of the city's climate change governance is the rise in privately controlled solar energy systems. SolarCity, a private solar company owned by Tesla, controls 70 percent of rooftop solar energy systems in Los Angeles as a result of participating in the city's Solar Incentive Program. The city does not currently have a residential feed-in tariff system, which would allow homeowners to install solar energy systems and sell it back to the city. Rather, SolarCity has developed relationships of its own with homeowners that allow them to install solar power systems and share the revenue with the homeowners. SolarCity's growing share of the local solar power market is seen as a deterrent to developing a residential feed-in tariff system, as an arrangement would have to be negotiated between LADWP and SolarCity. "It is essentially now a bottom-up utility, it controls a lot of the city's energy."³¹ In part to counter these concerns, the LADWP is developing a Community Solar Program through which customers can buy shares in solar energy projects installed on city-owned buildings and lock in their energy rates.

Going forward, Los Angeles has been confident in putting out additional GHG reduction goals because of the gains still to be had from moving away from coal. Coal represents 30 percent of the city's current energy mix, and Los Angeles has plans in place for withdrawing from all contracts for coal-generated power by 2025. The city council has its eye on renewable energy, recently passing legislation to study the options for moving toward 100 percent renewable energy sources. The state has also required the city get 50 percent of its energy supplies from renewable sources by 2030. Many in the city believe the transition process going forward will prove more challenging than the process to date. Significantly reducing GHG emissions also requires a transition away from natural gas, further efforts to build energy efficient buildings, and fulfilling the city's vision of a low-emission transportation system.

Toronto

As was the case in New York City, Toronto's efforts to reduce GHG emissions have changed internal processes and culture within city government: "It has changed the culture within the city so much. There was a guy who runs the building where salt trucks are stored. He wrote a memo to his boss asking why don't we have solar panels on the roof and so they went for it. It didn't need to go through council or anything, the business cases are normal operating procedure now."³² The sustainability lens that New York City was able to fix onto city planning and practices has also emerged in Toronto but through a bottom-up process. For example, Solid Waste Management Services now takes a "triple bottom line approach" that mirrors the economic, environmental, and equity goals of sustainability.³³

These changes have been slow to permeate beyond city government. Practices in the private sector are not being shaped significantly by city policies, and there is generally low awareness among this community of the city's aims to significantly reduce GHG emissions. This is likely due in part to the city's ongoing fiscal constraints (Slack and Cote 2014). Toronto's persistent revenue shortages and inability to raise property taxes or tap meaningful revenue sources at higher levels leaves many agencies struggling to fulfill their basic missions. The lack of clear signaling from City Hall further marginalizes climate change mitigation aims.

Toronto is unique among the three cases in having played an important role in shaping—if not driving—provincial climate change policies, particularly with regard to energy use in buildings. As described by an official from the provincial Ministry of Energy,

> Toronto can be the showcase, that's their role. Toronto has been a leader in demonstrating the validity of policy ideas. That way, rather than the province always having to look to California, which they do a lot, Toronto is already doing it and it's working so the province can make the case for it. We can convince provincial stakeholders (builders, etc.) that the sky won't fall.[34]

Some key examples are the city's leadership in initiating an energy use benchmarking program, demonstrating the benefits of energy efficiency retrofits in social housing buildings, and developing green building guidelines for new buildings. Toronto has played a similar role in providing leadership and examples for other Canadian cities. "Progress in Toronto leads to progress in other municipalities."[35] The city helped ICLEI develop its guidelines for mitigation planning for Canadian cities.

The city's Better Buildings Partnership and the existence and work of TAF have been nationally and internationally influential.[36] For example, London, England, developed the London Climate Change Agency, a municipal company modeled after TAF and designed to help the private sector reach energy efficiency goals. The Better Buildings Partnership and TAF go beyond municipal operations to target community-wide emissions, which many cities are still hesitant to take on.[37]

Toronto is considering new approaches for the next round of GHG emission reductions. Large projects such as methane capture and phasing out coal have allowed Toronto to meet its targets so far. City council recently instructed the Energy and Environment Division, as part of the *TransformTO* initiative, to revisit the city's climate change mitigation plans to determine whether the programs and policies in place are adequate to meet the goal of reducing emissions 80 percent by 2050. A report to council in November 2016 found that existing programs and policies "will need to aggressively expand . . . over the intervening 33 years to meet the 2050 goal established by Council" (City of Toronto 2016b). Many of the short-term strategies recommended to the city focus on expanding existing programs, such as the Better Buildings Partnership, the Home Energy Loan Program, and the retrofitting of social housing buildings. Like New York City and Los Angeles, there is a sense that "we have the right ideas, the right programs, we just need to scale them up."[38]

The report also points to a number of new directions needed in the city, including decentralized community energy planning, deployment of renewable energy, and enabling of electric vehicles. There a sense in the report, and among decision makers and stakeholders, that deeper changes to the city and its infra-

TABLE 5.1 Comparing the destinations of New York City, Los Angeles, and Toronto

	GHG EMISSION REDUCTIONS (ABSOLUTE)	GHG EMISSION REDUCTIONS (INTENSITY)	PRIMARY CONTRIBUTOR(S) TO EMISSION REDUCTIONS	BROADER IMPACTS
New York City	12% citywide^ 13% for city government	17% citywide^	Shift from coal to natural gas Methane capture	Governance norms, culture in city government Climate change mitigation policy and planning in other cities
Los Angeles	20% citywide* 25% for city government	28% citywide*	Shift from coal to renewable energy Methane capture	Transparency and accountability at the LADWP Enhanced leadership role of the mayor
Toronto	26% citywide# 19% for city government	31% citywide#	Shift from coal to nuclear, renewables Methane capture	Sustainability culture in city government Climate change mitigation policy at the provincial level

^relative to 2005 baseline
*relative to 1990 baseline
#relative to 2004 baseline

structure will be required to meet the larger targets. Some estimate that 1 million residential units need to be deeply retrofitted to meet the city's climate change goals, while just two hundred currently take advantage of city programs.[39] Transportation investments have so far maintained the modal split while accommodating growth, rather than facilitating real changes in the way people move throughout the city.[40]

Toronto will need to move beyond marginal change in these areas to keep pace with its goals. For this reason a greater emphasis is being placed on engaging and collaborating with stakeholders, including the city's energy utilities, on the development of a strategy for 2050. The *TransformTO* initiative is asking all of the right questions, but support from city council remains tenuous.

Proof of Principle

The cases provide initial evidence that urban climate change mitigation governance works. The three cities have played a direct or indirect role in reducing GHG emissions at a scale and pace that has kept them on track to meeting their longer-term goals. The effects of urban climate change governance also extend beyond measures of GHG emission reductions. The cities' efforts have helped to reshape urban governance norms, decision-making institutions, and political alliances. They have also led to change at other levels of government and in other cities. These findings show that what city governments do matters; they have the ability to initiate a process of redirecting their cities. While the notion that city governments matter has been central to the rhetoric surrounding cities and climate change, an emphasis on barriers and constraints had left us to wonder what cities might accomplish.

In doing the initial work of reducing GHG emissions, and beginning to explore what it means to mobilize resources and actors for climate change mitigation, these cities are developing new knowledge about what governing for transformative change requires: building institutions, coalitions, and capacities. The true test of urban climate change mitigation governance will come as these and other cities take on the task of moving from somewhat incremental to more transformative change. While increasing renewable energy by 20 percent or developing a city-scale building energy code is no small feat, each city faces a new challenge ahead. Current efforts must be scaled up dramatically to achieve an 80 percent reduction in urban GHG emissions. Moving from incremental to transformative change will require cities to grapple more seriously with broader city and regional politics, avenues for engaged and inclusive decision making, and the challenges of producing robust and accessible data infrastructure.

Conclusion

PROSPECTS AND CONSEQUENCES OF REPOWERING CITIES

> **When people are given the opportunity to steer government, it's their turn to make their mark.**
> *Jerrel Burney, former New York City Council staff member*

> **We're moving an aircraft carrier here.**
> *Rohit Aggarwala, former director of the New York City Office of Long Term Planning and Sustainability*

Local leadership has shifted the landscape of climate change mitigation, challenging us to confront the role of city governments in the governance of GHG emissions. Repowering cities physically to reduce emissions requires we also repower cities politically and institutionally to mobilize the resources and actors needed for such transformational change. Activating the governing capacities of city governments is the fundamental challenge of urban governance; this challenge is only amplified in the context of a complex and uncertain policy issue like climate change mitigation. In this book I examine the policies and governance strategies implicated by a locally led climate change mitigation agenda. The framework I develop for understanding the process of repowering cities provides a means for distinguishing analytically and empirically between the policy agendas city governments develop for reducing GHG emissions, the governing strategies they use to implement these agendas, and the direct and indirect ways cities contribute to climate change mitigation.

Urban climate change mitigation policy agendas are shaped by a city's political, economic, social, and technical context. City governments target sectors and sources of emissions that are contributing the most to their emissions inventory or represent easy wins or low-hanging fruit. They also choose governing modes that align with their institutional and political realities: relying on more or less direct intervention as opportunities present themselves. This flexibility and range of approaches to urban climate change mitigation allows cities in many different settings to tailor a policy agenda to their circumstances. The experiences of New York City, Los Angeles, and Toronto show that state policies can serve to constrain or enable certain policy choices but are not playing an outsized role in determining a city's agenda. Even in the case of Los Angeles, California's leadership on climate change has often been viewed as a floor for action rather than a ceiling.

City governments must mobilize the resources and actors needed to make significant changes to urban infrastructure, lifestyles, and services to pursue and implement climate change mitigation policy agendas. City governments must move from adopting a policy platform or emission reduction target to reconfiguring political, institutional, and organization landscapes. This task engages significant uncertainty and complexity. City governments can use the governing strategies of institution building, coalition building, and capacity building to overcome this uncertainty and steer in a new direction.

New York City, Los Angeles, and Toronto have each relied on these governing strategies. They have reorganized city departments and mayoral offices, incorporated new sources of expertise in decision-making processes, built and managed networks of stakeholders, leveraged and responded to electoral demands, and developed partnerships beyond their borders. The cities have often incorporated these governing strategies on an ad hoc basis: when political support became hard to come by, when public opposition surprisingly flared, or when data and capacity gaps revealed themselves. These governance strategies play an important role in mobilizing actors and resources regardless of the particular contours of a city's climate policy agenda. Despite differences in political leadership or state or provincial policy, all city governments have had to activate their own capacity to redirect and reconfigure—to *repower*—the cities they govern.

Perhaps the most critical question is what difference these efforts have made. These three cities have experienced demonstrable but modest reductions in GHG emissions; they are on track so far to meet their medium- and long-term goals. Linking these changes in GHG emissions to specific actions by the city governments remains challenging. In some cases, state or provincial actions played a large role, such as the decommissioning of coal-fired power plants in Ontario or the move to natural gas in New York state. Nonetheless, there are concrete examples of policies and programs in the cities that are working: Measure R in Los Angeles is generating real revenue for transit projects, New York City's Better Buildings program is reconfiguring how the city's buildings are powered and heated, and Toronto is capturing significant quantities of methane from its landfills. The challenge arises from the pace and scale at which these programs are operating.

The contributions these cities are making to climate change mitigation extend beyond any direct reductions in their own GHG emissions. Their leadership, learning, and experimentation are catalyzing change at other levels of government and in other cities. Their efforts to reduce GHG emissions are also reshaping the city in broader ways, leading to organization and institutional changes such as greater transparency of the LADWP, new inter-organizational relationships in Toronto, and a culture shift in New York City's business and real estate communities. The effects of urban climate change mitigation governance are as catalytic as they are atmospheric.

Looking beyond the Cases

New York City, Los Angeles, and Toronto are large, globally connected North American cities that have relatively high capacity for taking on a new problem like climate change. The findings and framework I put forward have relevance to other types of cities. The framework presupposes a commitment to address climate change; it aims to deconstruct the process of governing toward such a goal once it has been set. As more and more cities take up climate change mitigation as a policy aim, it is increasingly difficult to identify cities that are likely or unlikely to participate. Despite being located in one of the largest oil-producing regions of the United States, Houston, Texas, has set a goal of reaching 100 percent green power and recently built a 50 megawatt solar power facility. Small cities such as Grand Rapids, Michigan; Fayetteville, Arkansas; Anderson, South Carolina; and London, Ontario, have set GHG emission reductions targets and are charting their progress. Local climate action is not confined to North America and Europe: ten African cities, twelve Chinese cities, twelve Latin American cities, and eight south and west Asian cities are now members of the C40. The diversity of cities setting GHG emission reduction targets emphasizes the importance of shifting our focus to understanding what it takes for cities to make good on their pledges.

Climate change mitigation policy agendas are what we should expect to vary between cities. Reducing GHG emissions in a thriving metropolis requires a very different set of policy tools than reducing GHG emissions in a small or shrinking city. Strong city governments may benefit from replicating New York City's efforts to require energy conservation in privately owned buildings; city governments with municipal utilities may learn from the development of solar energy incentives by LADWP; city governments that lack authority or political backing can adopt facilitating and self-governing approaches similar to Toronto's. Cities operating in similar political or institutional contexts can learn from one another about developing climate change mitigation policy agendas that are aligned with these shared dimensions.

The central task of mobilizing actors and resources in order to govern is common to different types of cities. In fact, the task of activating a capacity to govern may be all the more crucial for city governments facing greater political or institutional challenges than the cases in this book. Coalitions may look different, capacity gaps may emerge in different ways, and state or federal governments may play different roles, but the need for city governments to build institutions, coalitions, and capacities in order to govern GHG emissions will remain the same. Fortunately, these are things city governments have more control over: while they cannot well control the policy choices made by federal governments, or go back in time to unbuild highways, they can make choices about how to best move forward.

As more and more cities are expressing an interest and a commitment to reducing GHG emissions, many are wondering where to start. The opportunities for learning between cities about *governance* are tremendous and, if tapped, could serve as an additional catalyst for progress on climate change. The experiences of these cities provide evidence that it is not just *what* city government do but *how* they do it that matters. Local decision makers are often presented with case studies or best practices that highlight the policy options and choices for reducing emissions, but these do not necessarily equip them with an understanding of how they might pursue such an initiative in their own city. City governments should be sharing and comparing experiences with the governing strategies: New York City's experiences with coalition building or institution building may be more useful to Grand Rapids or Accra than the technical details of their building code overhaul. Such learning and sharing can be facilitated by networks such as ICLEI and the C40: rather than highlighting a successful building energy use regulation, city networks could highlight a successful stakeholder advisory group or the strategies that led to a successful public referendum.

This framework can also support those in other levels of government or the nonprofit sector seeking to help and expand the efforts of city governments. Administrators and policymakers can act as coalition partners or help to frame interventions in ways that facilitate coalition building. Financial or regulatory programs should aim to support the development and deployment of capacity building and institutionalization alongside new incentives and innovations. Helping cities to navigate the *how* of governing GHG emissions is just as critical as helping them identify what interventions or technologies are available.

Toward Transformative Urban Climate Change Mitigation

New York City, Los Angeles, and Toronto find themselves at a critical juncture as they contemplate their long-term goal of reducing GHG emissions 80 percent by 2050. The low-hanging fruit has largely been picked, and each city is now contemplating ways of scaling up their efforts. The cities have often had success at the project scale, but to realize deep emission reductions many initiatives will need to be implemented citywide. Transportation has proven to be a particularly sticky challenge and will need to play a larger role in GHG emission reductions if the cities are to meet their long-term goals.

This challenge of "scaling up" is common in cities working to address climate change (Bulkeley 2013; Svara, Watt, and Jang 2013). Having taken the initial step of retrofitting some buildings, encouraging transit-oriented development in a changing neighborhood, or switching streetlights to LED bulbs, many cities are eager to have a wider impact. Incremental change seems to be a common product of urban climate change governance, but it is at odds with the ambitions cities are setting forth.

From the perspective of repowering cities, cities have not yet mobilized the resources commensurate with significant reductions in GHG emissions and are faced with the task of developing governing strategies that can deliver this scale of transformation. City governments have been able to use easy wins, pilot projects, and co-benefits to nudge their cities toward a low-emission future. But their ambitions extend far beyond what has been accomplished. Rather than a handful of leading corporations or neighborhoods, hundreds of publicly and privately owned buildings will need to be retrofitted for energy efficiency, renewable energy production will need to expand rapidly, and large swaths of the city will need to commute to and from work differently.

Reaching these goals will require an even deeper engagement with the remaining institutional, strategic, and substantive uncertainties of climate change mitigation. New York City and Toronto in particular have begun to identify paths toward 80 percent emission reductions through initiatives such as *New York City's Roadmap to 80x2050* and *TransformTO*. There are three ways the governing strategies central to repowering cities are being used in their pursuit of more significant GHG emission reductions: developing more participatory institutions, building capacity for climate "smart" decision making, and expanding and stabilizing climate coalitions. The ability of cities to fully develop these strategies may determine whether urban GHG emission reductions have stalled or are simply at an inflection point.

Building Participatory Institutions

Each city has built institutions that help to reduce uncertainty surrounding the decision-making process. Formalizing and codifying climate change mitigation goals and reporting requirements have been particularly important for maintaining and mobilizing resources and actors committed to climate change mitigation.

Going forward, investment in meaningful public engagement and inclusive decision-making processes will be necessary for more transformative urban climate change governance.

Public buy-in and support for climate change mitigation is considered necessary for pursuing deeper emission reductions. "Politics needs transformative ideas to be socialized first. The groundwork has to be laid."[1] Decision makers need political courage, which requires the communication and understanding of the climate change mitigation agenda and its benefits. "Politicians need to fear grassroots more than they fear the economic elite."[2] Some practitioners see these conversations as particularly critical in today's political climate—a way to counter a vision of fear and government ineptitude with one of livable cities and a new economy. Several compared the goal of transforming cities to significantly reduce GHG emissions to large-scale social projects of the past such as building the U.S. highway system or communicating the health effects of tobacco.

The need for this public support and buy-in for deep GHG emission reductions underscores the critical role of city governments. This is not an agenda that will be developed and implemented among elites. The cities are securing a constituency for climate change policy, but residents may still not view climate change mitigation as a core task for their city government.[3] Toronto is being particularly explicit in using broader engagement as a way of better mobilizing the resources needed for transformative change.[4] Los Angeles has already experienced some of the benefits and challenges of a mobilized and engaged public, with residents voting in favor of Measure R but challenging reforms and opacity at the LADWP.

Reaching longer-term climate change mitigation goals will require significant behavioral change. "We have to convince people who live in the city to do different things and do things differently."[5] Retrofitting a home to be more energy efficient or making different choices about commuting are not always changes people are eager to commit to. As one interviewee put it, "asking people to spend money on something that sucks is hard."[6] Greater public engagement has the potential to help cities identify *feasible* routes to deep GHG emission reductions. Residents can communicate their desires for a low-emission city and the pathways they think will lead to the city they want to live in in 2050. New York City is in the early stages of recognizing that this means going beyond elites and organized stakeholders and engaging the average New Yorker in conversations about climate change mitigation.[7]

The reality is that cities often fall short of participatory ideals in their work on climate change, as in other areas, and it is difficult to identify a "one size fits all" approach for cities to use (Hawkins and Wang 2012; Sarzynski 2015). Developing open, equitable, and transparent institutions to guide climate change decision making will be critical for reaching the goal of reducing GHG emissions 80 percent by 2050.

Capacity Building for Climate "Smart" Governance

The generation and maintenance of high-quality datasets to support climate change mitigation is a glaring gap in most cities. Tracking progress in reducing GHG emissions is challenging when cities use spotty and inconsistent accounting methods. City agencies, policymakers, stakeholders, and residents need tools for discerning program effectiveness and tracking their city's trajectory. New York City found that "programs with measurable benchmarks were most successful."[8] Without such tracking, there are very few opportunities to learn and adjust along the way, and ineffective or harmful programs are more likely to be allowed to continue. Decision makers in the three cities see data generation and availability as central to governing climate change going forward: "it is the most important thing for a city to do."[9]

Politics and administrative siloes often prevent greater urban GHG emissions data availability. Politically, data collection and sharing, and the enhanced transparency it brings, can be threatening to decision makers and managers. No one is eager to be revealed as a high-emitting city or department, and emissions accounting can provide unwelcome accountability. These political tensions are particularly apparent in Los Angeles, where struggles between the mayor, city council, and powerful agencies (including the LADWP but also the port and airport) have often led to reduced transparency. Navigating the political tensions and consequences of more and more readily available climate change mitigation data will be foundational to building capacity.

Administrative siloes and over-extended staff can inhibit the collection and maintenance of adequate data resources. This is evident in the cases and also in the challenges observed in the larger trend toward "smart cities" (Giest 2017). In Toronto, "everyone knows it's a problem, but we don't have the staff to take it on a lot of the time."[10] Staff are needed to collect and maintain datasets, analyze them, and share them in ways that are useful and appropriate. Staff may not have the expertise for such work or time available to devote to it. If data collection or sharing is not required, it is likely to not get done, especially if it is primarily to support another department.

One way of encouraging data availability is to educate decision makers and stakeholders on the value of data for policymaking. *TransformTO* is taking this approach with its modeling process and stakeholder advisory group. The group will be tasked with developing various future scenarios and will likely run up against data gaps that will prevent the full development of some scenarios. Stakeholders and decision makers will be able to see what would be possible to learn and understand if the data did exist, helping to demonstrate its usefulness as people explore climate change mitigation futures. In Los Angeles, the nonprofit Climate Resolve is working with partners to develop high-resolution models of the city able to create powerful visual images of the future and what good climate change policy could yield for the city and for residents. Demonstrating the value of high-quality, accessible climate change data can be an important step toward ensuring its production and availability.

Expanding and Stabilizing the Climate Change Coalition

Coalition building has been a key strategy for reducing strategic uncertainty and mobilizing needed resources for climate change mitigation. New York City has found it to be critical to its efforts to regulate energy use in buildings; Los Angeles has used it to build the support necessary for key initiatives like Measure R; Toronto was able to pass climate change legislation during a very conservative period in city politics by building a broad coalition around its energy conservation initiatives. It is hard to overstate the importance of political leadership in urban climate change governance, and this leadership is the product of electoral support and engaged, committed private and civil sector stakeholders. Going forward, decision makers and stakeholders need to build broader, more stable coalitions around climate change mitigation.

Labor, housing, and health organizations and advocates will be central to the formation of a broader climate coalition. The implications of deep GHG emission reductions for economic development and job creation need to be not only embedded in the political calculus but in the decision-making process itself. Job creation is often used as a metric for success and as a means of securing political support for climate change programs and legislation. As cities grapple with growing inequality and changing political economies, labor groups will be critical for mobilizing support and resources for climate change mitigation. Housing pressures and prices are high and rising in many North American cities. Energy efficiency retrofits and planning that prioritizes density and mixed-use development are climate change mitigation strategies that can also address housing needs. Framing climate change mitigation as a means to improve public health has proven effective for public and city council support but often occurs on a project-by-project basis. Truly building a coalition that includes public health advocates and stakeholders requires further engagement and collaboration with this community.

As part of building a broader climate change coalition, cities will need to grapple more explicitly with the justice and equity considerations of transformative reductions in GHG emissions. New York City, Los Angeles, and Toronto have each taken steps to build issues of inequality into the next phase of climate change mitigation policy and planning. While poorer communities are often most affected by the consequences of climate change (floods, heat, etc.), it may be the wealthiest who will see their lifestyles and neighborhoods change the most, as this is often the group responsible for a greater proportion of urban GHG emissions (Marcotullio et al. 2014). Programs like energy efficiency subsidies or community solar energy development must be accessible to all residents, but they often disproportionately reach wealthy and middle-class urban residents that arguably need less support in transitioning toward more climate-friendly lifestyles and behaviors (Reames 2016; Bednar, Reames, and Keoleian 2017). A more integrated and explicitly justice-oriented approach will be necessary to ensure that the transformations implied by an 80 percent emission reduction target are able to account for the heterogeneities of urban residents and their needs. These coalitions also need to expand beyond city borders. To make progress on deep GHG emission reductions, cities will need support from neighboring cities, county and regional bodies, and their state legislatures, who may also bring new sources of inequality and a greater heterogeneity of preferences and needs.

There are reasons to be skeptical of the effectiveness of such a "big tent" approach to climate change coalitions. With broad coalition building comes the risk of deprioritizing—or reprioritizing—GHG emission reductions as new members bring new interests. Broad coalition building may shift the focus away from climate change explicitly and introduce new political conflicts that would undermine the city's efforts. For example, introducing redistributive policies into a climate change program may galvanize greater resistance than might otherwise emerge.[11] Others question whether focusing on the "sweet spots in policy that provide people benefits" would actually be enough to meet the city's goals.[12] Los Angeles's experiences with reforming the LADWP highlight the importance of transparency to coalition building but also the potential limits to what the public is willing to support in the name of climate change mitigation. There may also be potential tensions and tradeoffs between public oversight and decision-making efficiency and capacities and a limit to what can be produced through consensual decision-making models. Relying on a business case for climate change mitigation may hide the need for the city to make tradeoffs, particularly in terms of large infrastructure investments.[13] As an advocate in Los Angeles points out, "sometimes you are fighting a zero sum fight and you intend to be."[14]

The process of coalition building for urban climate change governance may be the means by which cities discover the limits of their own abilities to reduce

GHG emissions. Moving toward deeper reductions raises the social and political stakes of climate change mitigation—alterations in governance institutions and the use and distribution of resources will intensify with the scale of intervention. As cities contemplate and work toward transformative change, tracking the broader implications in urban politics will be just as important as deciphering the trends and patterns in urban GHG emissions.

Confronting the Limits and Tradeoffs of a Local Response to Climate Change

Keeping global temperature increases below 2°C is a monumental task but critical to preventing the worst consequences of climate change (IPCC 2014a; Rockström et al. 2017). The ambitions and leadership of city governments for reducing GHG emissions have become increasingly central to global climate change solutions. The World Bank Group president recently included low-carbon cities as one of the top five ways to respond to climate change (Arkin 2015). The climate change commitments and ambitions of city governments are only increasing, and they have the potential to occupy an increasingly prominent place on many urban policy agendas. After analyzing the experiences of New York City, Los Angeles, and Toronto, I return to the questions that motivate this book: *Can* cities lead on climate change? Even if they can, *should* cities lead on climate change?

The scale of cities' contributions to climate change mitigation remains uncertain. New York City, Los Angeles, and Toronto have made noticeable but modest reductions to their GHG emissions—on the order of 15 percent below baseline levels. These three cities are working to increase the pace and scale of their efforts in ways that I described previously. There may be limits to such efforts, particularly in their ability to compel participation from a broader swath of local and regional actors. Expanding the aims of climate change mitigation efforts—from replacing light bulbs to fundamentally transforming the city—also expands the range of tradeoffs and uncertainties that city governments must confront.

The value of the framework I have laid out is that it provides an initial blueprint for understanding climate change mitigation as a policy task for city governments and the tools city governments have at their disposal for governing. It provides a set of working hypotheses about what it takes for city governments to set and pursue a transformative agenda. The direct and indirect contributions of urban climate change governance will be determined by how well, and in what ways, city governments are able to develop a strong policy agenda and leverage the strategies of institution building, coalition building, and capacity building.

Even if cities can chart a transformative path, what do we gain or lose from a locally led climate change mitigation agenda? Why burden our most financially constrained governments with one of the most critical challenges of our time? Is the decentralization of climate change mitigation actually welcomed by federal and state decision makers who can now more comfortably avoid confronting the wicked problem of reducing GHG emissions?

If cities are where climate change will be solved, it also means they will determine *how* it will be solved and who the winners and losers of climate change will be. The leadership of cities on climate change is often seen as heralding a more democratic, pragmatic agenda (e.g., B. Barber 2017). The experiences of New York City, Los Angeles, and Toronto refute the idea that shifting climate change policy to the local scale removes or mutes the politics of reducing GHG emissions. In fact, this is a deeply political project for cities and city governments. Institutional changes have made it mandatory for decision makers to account for GHG emissions and increased the transparency of the decision-making process following public skepticism. Coalition building aligns and mobilizes disparate political interests, stakeholders, and electorates to pursue a common goal; it often requires compromise and negotiation, and it can lead to unexpected outcomes. Capacity building requires overcoming political obstacles to data sharing or redirecting scarce resources to new programs. The novelty of cities acting on climate change is not in their pragmatic and apolitical decision making but rather their willingness to do the necessary political work.

A central role for cities in the transition to a low-carbon future creates a central role for urban politics. The democratic implications of a locally led climate change agenda are not immediately obvious. Calls for greater local autonomy and authority assume city governments will use their powers for good which, Jon Pierre points out, "is at best an empirical question which is yet to be tested systematically" (Pierre 2011, 110). Climate change mitigation policies can be regressive, placing the most cost on those least able to afford them and who do not contribute much to GHG emissions (Büchs, Bardsley, and Duwe 2011; McKendry 2016). Environmental agendas in cities can be captured by the liberal elite (Ghertner 2012; Goodling, Green, and McClintock 2015). Voter turnout is typically very low in North American cities, making it difficult to rely on traditional accountability and responsiveness mechanisms in the policy process. Reliance on informal authorities and city government's "power to" rather than "power over" in city government steering can also obscure decision-making processes and further complicate accountability measures (Nicholls 2005). The agenda could be hijacked, and city governments could ultimately steer in the "wrong" direction. Private finance stands to play an increasing role in local climate change initiatives, particularly in places like the United States where federal government funding has not been forthcoming. Growing intercity competition for recognition and leadership status on climate change mitigation may create incentives to reduce transparency. While I have argued that inclusive and transparent decision-making institutions are critical for cities to realize deep GHG emission reductions, most city governments are structured as classic Weberian bureaucracies and do not always adapt well to calls for greater transparency and inclusiveness (Aylett 2010).

Climate change mitigation can't be, and won't be, the sole priority for city governments. Cities are changing very rapidly. Many North American cities are experiencing demographic shifts that have implications for service provision, consumption patterns, and land use. "The city is moving fast, faster than the government."[15] Decision makers see people moving back to the downtown core and driving less, using bike lanes when they are built, and subsequently having different demands of their city governments. There is a "pent up desire to reclaim public space."[16] While movement back to the city may increase density and use of public and active transportation, gentrification processes in urban centers can also lead to conspicuous consumption (Carpenter and Lees 1995), increasing a city's GHG emissions. City governments will be grappling with these ongoing processes of change and renewal simultaneous to their efforts to steer toward climate change mitigation.

Repowering cities reshapes their institutions, political coalitions, and generative capacities. Cities begin their journey toward climate change mitigation with an existing set of social and structural inequalities, and it will be important to track the extent to which climate change mitigation "might serve to exacerbate or redress these underlying structural issues" (Bulkeley, Edwards, and Fuller 2014, 31). Our understanding of how cities can or should seek to reduce GHG emissions—the governance processes at play rather than the technical solutions available—has lagged significantly behind the ambitions of city governments. An important area for future research is how urban climate change governance interacts and intersects with the larger social and economic transitions many cities are experiencing.

Given the technical and political dimensions of meeting global mitigation targets, the work of city governments is necessary but not sufficient. Ultimately, cities will play a key role in what I see as a *locally mobilized polycentric system*. Decisions at multiple scales, by multiple actors, shape our collective emissions trajectory, and urban governance and policy is playing a critical mobilizing role. States and provinces like California and Quebec are implementing carbon cap and trade programs and continuing to move away from coal-generated electricity. The U.S. Clean Power Plan has yet to be dismantled entirely, and the Canadian government is developing a national cap and trade program. The private sector is also making moves toward a lower-carbon future, with Volvo announcing plans to move away from the combustion engine. Local action cannot replace these broader shifts—but it can support and facilitate them, and provide fertile ground for implementation. Likewise, the shifts these measures induce in energy and transportation systems will be experienced in cities, and their success and effectiveness may ultimately be determined by the social, political, and institutional capacities of city governments.

Repowering Cities for 21st Century Policy Challenges

The emergence of city leadership on climate change is one dimension of a larger urban inversion: city governments are tackling numerous novel and complex policy challenges that have historically been the purview of state or federal governments. The climate change ambitions of cities have helped spark new thinking about the role they can play in addressing pressing problems of the 21st century. Some see promise in the apparent pragmatism and nonpartisan nature of urban policymaking; others are encouraged by the liberal thinking and agglomeration economies found in cities. As city governments take on more complex policy challenges, new theoretical approaches are required for understanding this pursuit and its implications.

My framework for repowering cities is designed to unpack the challenge for city governments of pursuing a novel public aim in the context of a complex, uncertain, and dynamic governance landscape. The framework contributes to the growing body of work referred to as third generation governance studies (Skelcher, Sullivan, and Jeffares 2013, 15). Neil Bradford (2013) describes the aims of such approaches:

> Through contextualized comparisons of policies and places, third generation governance research tracks evolving relations and institutional experimentation as stakeholders perhaps once divided come to acknowledge their interdependence and pursue the material, moral, or civic benefits of joint work ... Municipal leaders try on new roles—catalyst, convener, facilitator, and partner—in the collective search for localized responses to complex policy challenges.

Repowering cities emphasizes and examines the role of city governments in these evolving relationships and develops criteria for successful urban responses to complex policy challenges.

The repowering cities framework has much to offer the study of other complex policy challenges city governments are now taking on. There is growing attention to social inequality in cities as demographic and economic shifts combine with growth-oriented policies (Hayward and Swanstrom 2011; Weaver 2016). Like climate change mitigation, reducing social inequality is a relatively new and immensely complex policy challenge for city governments to take on—it requires coordination among a range of sectors including housing, health, and transit; a coalition able to bring together adequate resources to address the problem; and new decision-making and monitoring capacities to understand how interventions can best address social inequality. Commentators often offer policy prescriptions such as increasing density and building transit, but this does little to help us understand *how* city governments might effectively take this challenge on and the political tensions that must be navigated. Repowering cities can provide a framework for understanding how city governments can or do pursue a social equality agenda, and it highlights the potential for both direct and indirect contributions from cities to the broader challenge of reducing social inequality.

The economic reinvention many cities are undertaking—particularly in the North American rust belt—is another example of a complex policy challenge where repowering cities can provide leverage. Cities are beginning to radically rethink their development strategies following decades of decline and population loss (Hackworth 2015, 2016; Neumann 2016). While an emphasis has often been placed on the importance of aligning development strategies with global or regional markets and opportunities to gain a competitive advantage, Beauregard and Pierre (2000) have cautioned that local capacity building, and tailoring initiatives to local political and organizational attributes, is a more prudent approach. David Imbroscio has long advocated for a transition to more locally based economies and place-based ownership models (Imbroscio 1997; Imbroscio, Williamson, and Alperovitz 2003). Repowering cities can be a means for understanding how such transitions might occur and what role city governments play in their pursuit. Using the repowering cities framework to understand such economic policy transitions in cities would be particularly fruitful theoretically because of the variation that is likely in the goals city governments are ultimately pursuing: economic reinvention is likely to mean very different things in different cities.

Far from the ungovernable city of the past, bold and ambitious city leaders are forging a new path for their cities. The task of repowering cities forces us to look more closely at city governments and their role in shaping and reshaping the city through the creation and pursuit of a bold public agenda. As Caroline Andrew argues, "city governments will not become more effective actors through provincial or federal recognition or power-sharing arrangements; rather, they can become more effective through *their* creation of more inclusive urban governance regimes" (Andrew 2001). This role for city governments has been under-studied, providing new opportunities to question the nature of urban politics and governance and the transferability and adaptability of urban governance theories; it creates opportunities to better support decision makers and stakeholders working to reach lofty goals. As cities become central to maintaining and forwarding a progressive agenda, understanding how city governments activate their capacity to govern, and the implications of their efforts both within the city and beyond, is more important than ever.

The progressive ambitions of city governments should be supported and celebrated, but we must also take seriously the tensions and contradictions they make visible. Repowering cities is not a function of further devolving authority or strengthening the formal powers of city governments. Rather, repowering cities will generate new urban institutions, coalitions, and capacities, and it will catalyze action in other places in ways that are not yet easily predictable. The choices city governments make have local and global consequences; understanding and fostering effective and equitable policy agendas and governance strategies can help us to close the gap between the rhetoric and reality of local leadership.

Acknowledgments

This book represents the culmination of several years of work, spanning multiple institutions and countries. I am deeply grateful for the support and engagement of so many talented and dedicated scholars and practitioners, and the ongoing encouragement and generosity of my family.

The willingness of people working in and around city government in New York City, Los Angeles, and Toronto to share their insights and experiences is what made this book possible. Addressing climate change typically represents a task added to an already overburdened workload, and is truly a monumental challenge; I deeply appreciate their willingness to take on the additional work of explaining it to a young professor. I am especially grateful to Rohit Aggarwala, Jonathan Parfrey, and Mary Pickering for participating in a public forum on cities and climate change at the University of Toronto. Hilary Firestone and David Jacot have also been particularly supportive of this work, for which I am grateful.

This book largely took shape while I was at the University of Toronto. I am fortunate to have had research support from a number of very talented and entrepreneurial students there. Matthew Lesch contributed to my background understanding of climate change policy development in New York City and Los Angeles, and, along with Nikhail Rai, helped collate greenhouse gas emissions data for the three cities. Laura Tozer, via the Environmental Governance Lab, helped with mitigation policies in Toronto, and Shona Zhang, on loan from UBC, helped me understand renewable energy development in the city of Toronto. Siddharth Chaudhari and Alexa Waud, while at Oxford, helped sort through the varying political dimensions of urban climate action. I also benefited greatly from presenting my work to colleagues at the University of Toronto through the Comparative Politics workshop, the Environmental Governance Lab's working paper series, and the Krause Memorial Lecture in the School of the Environment. The Institute for Municipal Finance and Governance and the Center for the Study of the United States helped this work reach a broader audience by encouraging me to distil the insights for decision makers, and providing a public forum for their presentation. I am particularly grateful for careful feedback from Matt Hoffmann, Theresa Enright, Richard Stren, Graham White, Steven Bernstein, Wendy Wong, and Antoinette Handley. Thank you to my patient officemate and excellent colleague, Stefan Renckens. I received financial support for the project from the University of Toronto through the Center for the Study of the United States, the Research and Scholarly Activity Fund, and the Connaught New Researcher Award.

I had the pleasure of presenting various aspects of this work to many groups outside the University of Toronto, and I am grateful for this opportunity and the feedback it generated. This includes presentations at the American Political Science Association, the Urban Affairs Association, the Environmental Politics and Governance conference in Bloomington, the International Workshop on Public Policy in Pittsburgh, and the City of Toronto. Sarah Anderson, Tom Koontz, Klaus Frey, Abby Williamson, Neil Bradford, Ramiro Berardo, Yogi Hale Hendlin, and Suso Baleato were especially generous with their engagement and feedback with sometimes very early chapter drafts.

The advantage of the lengthy process of writing a book is the opportunity it provides to engage and re-engage along the way. I am grateful for the generous ongoing collaboration and feedback from my practitioner colleagues in Toronto, particularly Mary Pickering, Linda Swanston, and David MacLeod.

Many terrific colleagues and mentors have shaped my thinking about cities and climate change. Working as a postdoc with Stephanie Pincetl at the University of California-Los Angeles allowed me to witness firsthand the challenges and opportunities of environmental policy research in and with large cities, terrain that she expertly navigates. As a postdoc at the National Center for Atmospheric Research, my work with Paty Romero-Lankao and the Climate Science and Applications Program introduced me to the challenging and exciting world of local climate change mitigation and adaptation. The international and interdisciplinary context of this work provided invaluable exposure to the intersecting realities of political and climatic forces in cities. At the U.S. Environmental Protection Agency my mentors, Dan Loughlin and Andy Miller, were supportive and innovative and encouraged me to explore my interests in creative ways. Along the way, I had the good fortune of engaging with Harriet Bulkeley, and the attendees of the cities and climate change early career workshop in Durham, particularly Eric Chu and Dave Gordon. Conversations with Megan Mullin, Richard Dilworth, Ron Vogel, and Matt Hoffmann have also been instrumental to my thinking.

My heartfelt thanks go to Michael McGandy at Cornell University Press. His support, humor, and professionalism were at times very needed salves and helped to bring out the best in my work. Thank you also to Karen Hwa and Michelle Witkowski for their expert editing and fielding of myriad questions. Thank you to Rachel Lyon for creating the index to this book.

In the course of undertaking the research and writing of this book our family doubled in size. Thank you to Pascal and Josephine for coming along, both literally by joining me for fieldwork, and metaphorically by being flexible, brave, and sometimes even patient. Thank you to my mother, Linda Russell, for being the linchpin of so many of our endeavors and grandmother extraordinaire. I have always had the infinite support of my father, Mark Hughes, and siblings, Paul, Amy and Eric, and this is at the root of any of my accomplishments. My mother and father in-law, Daniel and Frances, are perfect models of loving and supportive second parents, and I am so, so fortunate. Thank you to our extended family, including Lizzy and Eric, to whom I can now say definitively: "Yes, the book is finished." Thank you does not seem to be quite enough for Jean-Luc, my partner in all things. But thank you always.

Notes

NOTE ON INTERVIEW PROTOCOL

The interviews that inform this research and its conclusions were conducted with people working in and around the city governments of New York City, Los Angeles, and Toronto. In many cases, they were still actively engaged in climate change policy and politics in their cities. To ensure confidentiality, I have omitted their names and do not attribute quotes explicitly to an interviewee unless I received their permission to do so. When I do reference material that comes from these interviews, I include the date and location of the conversation.

INTRODUCTION

1. In-person interview, Toronto, June 22, 2016.

1. PROGRESS OR PIPE DREAM?

1. Formerly the Cities for Climate Protection program, or CCP; http://www.iclei.org/activities/agendas/low-carbon-city/gcc.html.

2. There is some debate surrounding the exact proportion of global emissions that can be attributed to cities (Satterthwaite 2008). There is a growing consensus that cities are significant sources of GHG emissions and that this contribution will increase as urban populations continue to grow.

3. Mitigation can be accomplished either by reducing GHG emissions from human activities or physically removing GHGs from the atmosphere, but the focus of climate change mitigation policy has largely been on the former approach.

4. This was a common approach in Canada. For example, in 2003 FCM delegates in Canada representing 290 local governments endorsed a Municipal Leaders Resolution on Climate Change, which established a Kyoto-like target of reducing GHG emissions 6 percent below baseline in ten years (Federation of Canadian Municipalities 2015a).

5. The diffuse nature of city climate change action means that precise estimates of the scale of city involvement are challenging. Participation in intercity networks, such as the C40 Cities Climate Leadership Group (C40) and ICLEI's Green Climate Cities, provides a useful proxy for city actions. The C40 formed in 2005 and has grown rapidly from twenty initial members to more than eighty large cities around the world. The network today represents more than 600 million people, and 25 percent of global GDP (C40 2016). Member cities are encouraged to set a GHG emissions reduction target of 80 percent by 2050 and have committed to the Paris Agreement's goal of limiting global temperature increase to 1.5°C. The C40 reports that the commitments made by member cities would reduce global GHG emissions by around 6 percent (C40 2016). The GCC is also a global network of cities, growing from fourteen initial members in 1993 to well over one thousand members from eighty-six countries and headquarters in the United States, Canada, Europe, Asia, Africa, and Australia. A recent survey of U.S. municipal governments found that over two hundred have set GHG emission reduction targets (Svara, Read, and Moulder 2011).

6. In-person interview, Toronto, October 27, 2015; and Rohit Aggarwala, in remarks at the University of Toronto, March 27, 2017.

7. In-person interview, Toronto, October 27, 2015.

2. EVALUATING URBAN GOVERNANCE

1. In-person interview, Los Angeles, August 19, 2015.
2. Jacobs (2011) similarly argues that there are three hurdles to future-oriented policymaking: problems of electoral risk, problems of predicting long-term policy effects, and problems of institutional capacity.
3. Telephone interview, Los Angeles, September 10, 2015.
4. See, for example, Kennedy et al. (2015). Emission inventories were completed somewhat independently from city governments but are very difficult to compare between cities and represent different time periods in each city.
5. Or not (see C. D. Gore 2010).

3. MADE TO MEASURE

1. These relative contributions to GHG emissions have remained constant since 1995.
2. In-person interview, New York City, June 30, 2015.
3. In-person interview, New York City, June 30, 2015.
4. While ConEdison is still tasked with electricity delivery and grid maintenance, electricity generation is now controlled by five distinct entities: ConEdison, Reliant Resources, NRG Energy, KeySpan Energy, and the New York Power Authority (a state agency).
5. In-person interview, New York City, July 7, 2015.
6. In-person interview, New York City, June 30, 2015.
7. In-person interview, New York City, July 1, 2015. The mayor of New York City has the power to appoint and remove heads of the city's more than fifty departments and agencies. The mayor is responsible for policy implementation and so has full authority to restructure and redistribute administrative resources to facilitate his or her agenda (Berg 2007, 180). The mayor is also responsible for introducing a multibillion-dollar budget for the city each year, is able to introduce legislation, and has veto power over legislative actions.
8. In-person interview, New York City, July 2, 2015.
9. Rohit Aggarwala, in remarks made at the University of Toronto, March 27, 2017. While the city was severely in debt in 2001, to the point where the state was threatening to take over, the Bloomberg administration introduced spending cuts and a large property tax increase (18.5 percent in 2002), and by 2006 the city had generally recovered.
10. In-person interview, New York City, June 30, 2015.
11. In-person interview, New York City, July 2, 2015.
12. In-person interview, New York City, July 1, 2015.
13. In-person interview, Los Angeles, August 19, 2015; in-person interview, Los Angeles, August 26, 2015.
14. Telephone interview, Los Angeles, March 3, 2016.
15. In-person interview, Los Angeles, August 18, 2015.
16. In-person interview, Los Angeles, August 17, 2015; in-person interview, August 26, 2015.
17. In-person interview, Los Angeles, August 28, 2015.
18. In-person interview, Los Angeles, August 26, 2015.
19. In-person interview, Los Angeles, August 19, 2015.
20. E-mails, Toronto, May 23, 2017.
21. In-person interview, Toronto, June 23, 2016.
22. In-person interview, Toronto, May 13, 2016.
23. In-person interview, Toronto, October 28, 2015.
24. Peter Wallace, in remarks made at the University of Toronto, October 14, 2015.
25. In-person interview, Toronto, June 22, 2016.

26. In-person interview, Toronto, June 22, 2016.
27. In-person interview, Toronto, May 17, 2016; in-person interview, Toronto, April 8, 2016.
28. In-person interview, New York City, July 7, 2015. Also, for a comprehensive review of the process of creating *PlaNYC*, see ICLEI 2010.
29. In-person interview, New York City, July 7, 2015.
30. In-person interview, New York City, July 6, 2015.
31. In-person interview, New York City, July 9, 2015.
32. In-person interview, New York City, July 7, 2015.
33. In-person interview, New York City, July 2, 2015.
34. In-person interview, New York City, July 9, 2015.
35. In-person interview, New York City, June 29, 2015.
36. Rohit Aggarwala, in remarks made at the University of Toronto, March 27, 2017.
37. In-person interview, New York City, July 7, 2015.
38. In-person interviews, New York City, June 29, 2015, and July 7, 2015.
39. In-person interview, New York City, July 9, 2015.
40. In-person interview, New York City, June 29, 2015.
41. In-person interview, Los Angeles, August 26, 2015.
42. Villaraigosa lost the election to James Hahn, despite winning the city's nonpartisan primary election. Hahn was able to capture a large portion of African American voters (Wright and Middleton 2001). Villaraigosa ran again in 2005 and defeated Hahn with a broad coalition of Latinos, African Americans, and white liberals.
43. In-person interview, Los Angeles, August 20, 2015.
44. The group "included mainstream environmental organizations such as the Natural Resources Defense Council, Friends of the LA River, and Heal the Bay; environmental justice organizations such as Communities for a Better Environment, East Yard Communities for Environmental Justice, and Pacoima Beautiful; constituent-based advocacy organizations such as Physicians for Social Responsibility-LA; and community organizing groups such as ACORN and Union de Vecinos" (Matsuoka and Gottlieb 2013, 449).
45. State legislation (SB 1) in 2007 required municipal utilities to contribute to a statewide goal of 3,000 MW of solar energy production. The Obama administration provided 30 percent federal tax rebates for solar energy projects. See also Villaraigosa, Sivaram, and Nichols 2013.
46. In-person interview, Los Angeles, August 19, 2015.
47. In-person interview, Los Angeles, August 26, 2015.
48. Incentive levels were high in part because LADWP's energy rates are quite low, lower than other similar utilities in California. This is attractive to ratepayers but means that solar energy systems have a longer payoff time, which can discourage participation in new programs.
49. In-person interview, Los Angeles, August 26, 2015.
50. In-person interview, Los Angeles, August 19, 2015.
51. In-person interview, Los Angeles, August 17, 2015.
52. In-person interview, Los Angeles, August 18, 2015.
53. In-person interview, Toronto, May 5, 2016.
54. In-person interview, Toronto, May 5, 2016.
55. In-person interview, Toronto, October 27, 2015.
56. In-person interview, Toronto, May 5, 2016.
57. In-person interviews, Toronto, April 8, 2016; October 28, 2015; June 22, 2016.
58. In-person interview, Toronto, November 4, 2015.
59. In-person interview, Toronto, October 27, 2015.
60. In-person interview, Toronto, October 27, 2015.

61. In-person interview, Toronto, November 4, 2015.
62. In-person interview, Toronto, April 8, 2016.
63. In-person interview, Toronto, April 8, 2016.
64. In-person interview, Toronto, May 17, 2016.
65. In-person interview, Toronto, May 17, 2016.
66. In-person interview, Toronto, October 27, 2015.
67. In-person interview, Toronto, June 7, 2016.
68. In-person interview, Toronto, November 5, 2015.
69. In-person interview, Toronto, November 5, 2015.
70. In-person interview, Toronto, November 7, 2016.
71. In-person interview, Toronto, July 6, 2016.
72. In-person interview, Toronto, July 6, 2016.
73. Telephone interview, Los Angeles, August 31, 2015.

4. THE MEANS BEHIND THE METHODS

1. In-person interview, David Jacot, Los Angeles, August 17, 2015.
2. Rohit Aggarwala, in remarks made at the University of Toronto, March 27, 2017.
3. In-person interview, New York City, July 2, 2015.
4. In-person interview, New York City, July 6, 2015.
5. In-person interview, New York City, July 9, 2015.
6. In-person interview, New York City, July 6, 2017.
7. In-person interview, New York City, July 6, 2017.
8. Rohit Aggarwala, in remarks made at the University of Toronto, March 27, 2017.
9. In-person interview, Los Angeles, August 18, 2015.
10. In-person interview, Los Angeles, August 20, 2015.
11. In-person interview, Los Angeles, August 17, 2015.
12. In-person interview, Toronto, October 27, 2015.
13. In-person interview, Toronto, May 5, 2016.
14. In-person interview, Toronto, October 27, 2015.
15. In-person interview, Toronto, May 13, 2016.
16. In-person interview, Toronto, June 22, 2016.
17. In-person interview, New York City, July 6, 2015.
18. In-person interview, New York City, July 9, 2015.
19. In-person interview, New York City, July 7, 2015.
20. In-person interview, Toronto, April 8, 2016.
21. *Green LA* was presented as a living document that would be revised and expanded as the city began the (unspecified and never fully realized) process of soliciting input from the public.
22. In-person interview, New York City, July 7, 2015.
23. In-person interview, New York City, July 6, 2015.
24. In-person interview, New York City, July 7, 2015.
25. In-person interview, New York City, July 7, 2015.
26. In-person interview, New York City, July 7, 2015.
27. In-person interview, New York City, July 9, 2015.
28. In-person interview, Los Angeles, August 19, 2015.
29. In-person interview, Los Angeles, August 19, 2015.
30. In-person interview, Los Angeles, August 17, 2015.
31. In-person interview, Los Angeles, August 17, 2015.
32. In-person interview, Toronto, May 17, 2016.
33. In-person interview, Toronto, November 11, 2016.
34. In-person interview, Toronto, July 6, 2016.

35. In-person interview, Los Angeles, August 19, 2015.
36. In-person interview, Los Angeles, August 26, 2015.
37. In-person interview, Toronto, June 22, 2016.
38. In-person interview, Toronto, June 22, 2016.
39. In-person interview, Toronto, June 22, 2016.
40. Rohit Aggarwala, in remarks made at the University of Toronto, March 27, 2017.
41. In-person interview, New York City, July 2, 2015.
42. In-person interview, New York City, July 7, 2015.
43. In-person interview, New York City, July 7, 2015.
44. The city hoped Portfolio Manager would increase compliance because it doesn't require manual entry and building owners can directly upload their energy use information. There was an additional benefit to the city from working closely with the EPA, who integrated the city's feedback into the Portfolio Manager tool.
45. Hillary Firestone, in a presentation given in Los Angeles to the Green Building Council of the San Fernando Valley, August 17, 2015.
46. In-person interview, Toronto, June 22, 2016.
47. In-person interview, Los Angeles, August 17, 2015.
48. In-person interview, Los Angeles, August 17, 2015.
49. In-person interview, Los Angeles, August 19, 2015.
50. In-person interview, Los Angeles, August 17, 2015.
51. In-person interview, Los Angeles, August 17, 2015.
52. In-person interview, Toronto, November 5, 2015.
53. In-person interview, Toronto, June 22, 2016.
54. In-person interview, Toronto, May 13, 2016.
55. In-person interview, New York City, July 9, 2015.
56. In-person interview, New York City, July 6, 2015.
57. In-person interview, New York City, July 2, 2015.
58. In-person interview, New York City, July 9, 2015.

5. ARE WE THERE YET?

1. While originally the city was using ICLEI's U.S. Community Protocol for Accounting and Reporting of GHG Emissions, in 2016 they, like many large cities, switched to using the Global Protocol for Community-Scale GHG Emission Inventory (GPC), jointly produced by ICLEI, the World Resources Institute, and C40 cities. New York City uses the GPC-BASIC inventory, which includes "direct and indirect emissions from energy used by buildings and other stationary sources; on-road transportation and public transit within the geographic borders of New York City; and fugitive GHG emissions from wastewater treatment, in city landfills, solid waste disposed out of the city, and natural gas distribution within New York City" (City of New York 2014, 13).
2. In-person interview, New York City, June 29, 2015.
3. In-person interviews, New York City, July 9, 2015, and July 6, 2015.
4. In-person interview, New York City, July 9, 2015.
5. Rohit Aggarwala, in remarks made at the University of Toronto, March 27, 2017.
6. In-person interview, New York City, July 9, 2015.
7. In-person interview, New York City, July 1, 2015.
8. In-person interview, New York City, July 1, 2015.
9. Telephone interview, Los Angeles, September 10, 2015.
10. In-person interview, Los Angeles, August 18, 2015.
11. In-person interview, Los Angeles, August 18, 2015.
12. While the city conducted one additional GHG inventory in 2007, it used a different methodology and is therefore not comparable.

13. In-person interviews, Los Angeles, August 26, 2015, and August 18, 2015.
14. In-person interview, Los Angeles, August 25, 2015.
15. In-person interview, Toronto, June 22, 2016.
16. In-person interview, Toronto, November 11, 2016.
17. Prior to the coal phase-out the deep lake water cooling project had reduced Toronto's GHG emissions by 600,000 tons. Today, there is no impact on GHG emissions from the project.
18. In-person interviews, Toronto, April 8, 2016, and November 11, 2016.
19. In-person interview, Toronto, April 8, 2016.
20. In-person interview, Toronto, November 5, 2015.
21. In-person interview, New York City, July 1, 2015.
22. In-person interview, New York City, July 6, 2015.
23. In-person interview, New York City, July 7, 2015.
24. In-person interview, New York City, July 9, 2015.
25. In-person interview, New York City, July 9, 2015.
26. In-person interview, Los Angeles, August 17, 2015.
27. In-person interview, Los Angeles, August 19, 2015.
28. In-person interview, Los Angeles, August 17, 2015.
29. In-person interview, Los Angles, August 19, 2015.
30. In-person interview, Los Angeles, August 28, 2015.
31. In-person interview, Los Angeles, August 26, 2015.
32. In-person interview, Toronto, July 6, 2016.
33. In-person interview, Toronto, June 7, 2016.
34. In-person interview, Toronto, June 23, 2016.
35. In-person interview, Toronto, October 28, 2016.
36. In-person interview, Toronto, May 17, 2016.
37. In-person interview, Toronto, May 17, 2016.
38. In-person interview, Toronto, November 11, 2016.
39. In-person interview, Toronto, July 6, 2016.
40. In-person interview, Toronto, November 11, 2016.

CONCLUSION

1. Rohit Aggarwala, in remarks made at the University of Toronto, March 27, 2017.
2. In-person interview, Toronto, May 5, 2016.
3. In-person interview, Toronto, May 13, 2016.
4. In-person interview, Toronto, June 22, 2016.
5. In-person interview, Toronto, April 8, 2016.
6. In-person interview, New York City, July 1, 2015.
7. In-person interview, Los Angeles, August 17, 2015.
8. In-person interview, New York City, July 6, 2015.
9. Telephone interview, Los Angeles, March 3, 2016.
10. In-person interview, New York City, June 22, 2015.
11. Rohit Aggarwala, in remarks made at the University of Toronto, March 27, 2017.
12. In-person interview, Los Angeles, August 26, 2015.
13. In-person interview, Toronto, October 28, 2015.
14. In-person interview, Los Angeles, August 26, 2015.
15. In-person interview, Toronto, April 8, 2016.
16. In-person interview, Los Angeles, August 26, 2015.

References

American Road & Transportation Builders Association. 2013. "Vehicle Miles Traveled (VMT) & Congestion Pricing Programs." http://www.transportationinvestment.org/wp-content/uploads/2014/01/Congestion-Pricing-Overview-VMT-Cordon-Pricing.pdf.

Andrew, Caroline. 2001. "The Shame of (Ignoring) the Cities." *Journal of Canadian Studies* 35 (4): 100–110.

Andrew, Caroline, Katherine A. Graham, and Susan D. Phillips, eds. 2002. *Urban Affairs: Back on the Policy Agenda*. Montreal & Kingston: McGill-Queen's University Press.

Anguelovski, Isabelle, and JoAnn Carmin. 2011. "Something Borrowed, Everything New: Innovation and Institutionalization in Urban Climate Governance." *Current Opinion in Environmental Sustainability* 3 (3): 169–75. https://doi.org/10.1016/j.cosust.2010.12.017.

Arkin, Benjamin. 2015. "5 Ways to Reduce the Drivers of Climate Change." World Bank.

ARUP. 2014. "Climate Action in Megacities: C40 Cities Baseline and Opportunities Volume 2.0." London, UK. http://c40.org/blog_posts/CAM2.

———. 2015. "Climate Action in Megacities 3.0." London, UK. http://www.cam3.c40.org/images/C40ClimateActionInMegacities3.pdf.

Aylett, Alex. 2010. "Conflict, Collaboration and Climate Change: Participatory Democracy and Urban Environmental Struggles in Durban, South Africa." *International Journal of Urban and Regional Research* 34 (3): 478–95. https://doi.org/10.1111/j.1468-2427.2010.00964.x.

———. 2013. "The Socio-Institutional Dynamics of Urban Climate Governance: A Comparative Analysis of Innovation and Change in Durban (KZN, South Africa) and Portland (OR, USA)." *Urban Studies* 50 (7): 1386–1402. https://doi.org/10.1177/0042098013480968.

———. 2014. "Progress and Challenges in the Urban Governance of Climate Change: Results of a Global Survey." Cambridge, MA: Massachusetts Institute of Technology, Department of Urban Studies and Planning.

Bagley, Katherine, and Maria Gallucci. 2013. *Bloomberg's Hidden Legacy: Climate Change and the Future of New York City*. InsideClimate News. https://insideclimatenews.org/content/bloombergs-hidden-legacy.

Barber, Benjamin. 2017. *Cool Cities: Urban Sovereignty and the Fix for Global Warming*. New Haven, CT: Yale University Press.

Barber, Benjamin R. 2013. *If Mayors Ruled the World: Dysfunctional Nations, Rising Cities*. New Haven, CT: Yale University Press.

Barr, Stewart, and Jan Prillwitz. 2014. "A Smarter Choice? Exploring the Behaviour Change Agenda for Environmentally Sustainable Mobility." *Environment and Planning C: Government and Policy* 32 (1): 1–19.

Barrett, Beth. 2007. "More than 13% of DWP Workers Are Paid $100,000 and Up." *Los Angeles Daily News*, September 30, 2007. http://www.dailynews.com/article/zz/20070930/NEWS/709309888.

Bättig, Michèle B., and Thomas Bernauer. 2009. "National Institutions and Global Public Goods: Are Democracies More Cooperative in Climate Change Policy?" *International Organization*, 281–308.

Beal, Vincent, and Gilles Pinson. 2013. "When Mayors Go Global: International Strategies, Urban Governance and Leadership." *International Journal of Urban and Regional Research*. https://doi.org/10.1111/1468-2427.12018.

Beauregard, Robert A., and Jon Pierre. 2000. "Disputing the Global: A Sceptical View of Locality-Based International Initiatives." *Policy & Politics* 28 (4): 465–478. https://doi.org/10.1332/0305573002501081.

Bednar, Dominic J., Tony Gerard Reames, and Gregory A. Keoleian. 2017. "The Intersection of Energy and Justice: Modeling the Spatial, Racial/Ethnic and Socioeconomic Patterns of Urban Residential Heating Consumption and Efficiency in Detroit, Michigan." *Energy and Buildings* 143: 25–34.

Benize, Robert. 2017. "Kathleen Wynne Stopping John Tory's Plan for Tolls on DVP, Gardiner." *Star*, January 26, 2017. https://www.thestar.com/news/queenspark/2017/01/26/kathleen-wynne-stopping-john-torys-plan-for-tolls-on-dvp-gardiner.html.

Berg, Bruce. 2007. *New York City Politics: Governing Gotham*. Rutgers, NJ: Rutgers University Press.

Betsill, Michele M. 2001. "Mitigating Climate Change in US Cities: Opportunities and Obstacles." *Local Environment* 6 (4): 393–406. https://doi.org/10.1080/13549830120091699.

Betsill, Michele M., and Harriet Bulkeley. 2007. "Looking Back and Thinking Ahead: A Decade of Cities and Climate Change Research." *Local Environment* 12 (5): 447–456. https://doi.org/10.1080/13549830701659683.

Bierbaum, Rosina, Joel B. Smith, Arthur Lee, Maria Blair, Lynne Carter, F. Stuart Chapin III, Paul Fleming, et al. 2013. "A Comprehensive Review of Climate Adaptation in the United States: More Than Before, but Less Than Needed." *Mitigation and Adaptation Strategies for Global Change* 18 (3): 361–406.

Blakeley, E., and N. Green Leigh. 2010. *Planning Local Economic Development: Theory and Practice*. Thousand Oaks, CA: Sage.

Bohn, Sarah, and Eric Schiff. 2011. "The Great Recession and Distribution of Income in California." Oakland: Public Policy Institute of California.

Borger, Julian. 2001. "Bush Kills Global Warming Treaty." *Guardian*, 2001, March 29 edition. https://www.theguardian.com/environment/2001/mar/29/globalwarming.usnews.

Borick, Christopher P., Erick Lachapelle, and Barry G. Rabe. 2011. "Climate Compared: Public Opinion on Climate Change in the United States and Canada." *Issues in Governance Studies*, no. 39.

Boswell, Michael R., Adrienne I. Greve, and Tammy L. Seale. 2012. *Local Climate Action Planning*. Washington, DC: Island Press.

Bradford, Neil. 2013. "Urban Governance Hybrids for Complex Policy Initiatives: The New Localism in Toronto." Chicago, IL. Presented at the Annual Meeting of the American Political Science Association.

Bramwell, Allison, and Jon Pierre. 2013. "The Collaborative City: A New Mode of Urban Governance?" Chicago, IL. Presented at the Annual Meeting of the American Political Science Association.

Brody, Samuel D., Sammy Zahran, Himanshu Grover, and Arnold Vedlitz. 2008. "A Spatial Analysis of Local Climate Change Policy in the United States: Risk, Stress, and Opportunity." *Landscape and Urban Planning* 87 (1): 33–41. https://doi.org/10.1016/j.landurbplan.2008.04.003.

Büchs, Milena, Nicholas Bardsley, and Sebastian Duwe. 2011. "Who Bears the Brunt? Distributional Effects of Climate Change Mitigation Policies." *Critical Social Policy* 31 (2): 285–307.
Bulkeley, Harriet. 2010. "Cities and the Governing of Climate Change." *Annual Review of Environment and Resources* 35 (1): 229–253. https://doi.org/10.1146/annurev-environ-072809-101747.
———. 2013. *Cities and Climate Change*. New York: Routledge.
———. 2015. "Can Cities Realise Their Climate Potential? Reflections on COP21 Paris and Beyond." *Local Environment* 20 (11): 1405–1409. https://doi.org/10.1080/13549839.2015.1108715.
Bulkeley, Harriet, Liliana B. Andonova, Michele M. Betsill, Daniel Compagnon, Thomas Hale, Matthew J. Hoffmann, Peter Newell, Matthew Paterson, Charles Roger, and Stacy D. VanDeveer. 2014. *Transnational Climate Change Governance*. Cambridge: Cambridge University Press.
Bulkeley, Harriet, and Michele M. Betsill. 2003. *Cities and Climate Change: Urban Sustainability and Global Environmental Governance*. London and New York: Routledge.
———. 2013. "Revisiting the Urban Politics of Climate Change." *Environmental Politics* 22 (1): 136–154. https://doi.org/10.1080/09644016.2013.755797.
Bulkeley, Harriet, JoAnn Carmin, Vanesa Castán Broto, Gareth A. S. Edwards, and Sara Fuller. 2013. "Climate Justice and Global Cities: Mapping the Emerging Discourses." *Global Environmental Change* 23 (5): 914–925. https://doi.org/10.1016/j.gloenvcha.2013.05.010.
Bulkeley, Harriet, and Vanesa Castán Broto. 2013. "Government by Experiment? Global Cities and the Governing of Climate Change." *Transactions of the Institute of British Geographers* 38 (3): 361–375.
Bulkeley, Harriet, Vanesa Castán Broto, and Gareth A. S. Edwards. 2014. *An Urban Politics of Climate Change: Experimentation and the Governing of Socio-Technical Transitions*. London and New York: Routledge.
Bulkeley, Harriet, Gareth A. S. Edwards, and Sara Fuller. 2014. "Contesting Climate Justice in the City: Examining Politics and Practice in Urban Climate Change Experiments." *Global Environmental Change* 25 (March): 31–40. https://doi.org/10.1016/j.gloenvcha.2014.01.009.
Bulkeley, Harriet, and Kristine Kern. 2006. "Local Government and the Governing of Climate Change in Germany and the UK." *Urban Studies* 43 (12): 237–259.
Bulkeley, Harriet, Matt Watson, and Ray Hudson. 2007. "Modes of Governing Municipal Waste." *Environment and Planning A* 39: 2733–2753.
Burch, Sarah. 2010. "Transforming Barriers into Enablers of Action on Climate Change: Insights from Three Municipal Case Studies in British Columbia, Canada." *Global Environmental Change* 20 (2): 287–297.
Burch, Sarah, Alison Shaw, Ann Dale, and John Robinson. 2014. "Triggering Transformative Change: A Development Path Approach to Climate Change Response in Communities." *Climate Policy* 14 (4): 467–487. https://doi.org/10.1080/14693062.2014.876342.
Butler, Elizabeth R., and William H. Labhart. 2010. "Landmark Legislation Designed to Make New York City Greener." *Journal of Multistate Taxation and Incentives* 20 (8): 41–43.
C40. 2012a. "Case Study: Green Codes Task Force." C40. 2012. http://www.c40.org/case_studies/green-codes-task-force.
———. 2012b. "Trash to Cash: Methane Capture Generates $3–4 Million Annually." http://www.c40.org/case_studies/trash-to-cash-methane-capture-generates-3-4-million-annually.

———. 2016. "C40 by the Numbers." C40 Cities Climate Leadership Group. http://c40-production-images.s3.amazonaws.com/fact_sheets/images/10_C40_By_the_Numbers_April_2016.original.pdf?1459877919.

California Current. 2007. "LADWP's Rate Hike Plan Hits Resistance in City Council." *California Current*, November 2, 2007. http://cacurrent.com/subscriber/archives/20289.

California Energy Commission. 2007. "California's Residential Electricity Consumption, Prices, and Bills, 1980–2005." Staff Report CEC-200-2007-018. Sacramento, CA.

Carpenter, Juliet, and Loretta Lees. 1995. "Gentrification in New York, London and Paris: An International Comparison*." *International Journal of Urban and Regional Research* 19 (2): 286–303. https://doi.org/10.1111/j.1468-2427.1995.tb00505.x.

Castán Broto, Vanesa. 2017. "Urban Governance and the Politics of Climate Change." *World Development* 93: 1–15. https://doi.org/10.1016/j.worlddev.2016.12.031.

Castán Broto, Vanesa, and Harriet Bulkeley. 2013. "A Survey of Urban Climate Change Experiments in 100 Cities." *Global Environmental Change* 23: 92–102.

Center for Clean Air Policy. 2009. "Ask the Climate Question: Adapting to Climate Change Impacts in Urban Regions." Washington, DC: Urban Leaders Adaptation Initiative.

Chan, Sewell. 2007. "Is Manhattan Turning against Congestion Pricing?" *New York Times*, November 19. https://cityroom.blogs.nytimes.com/2007/11/19/is-manhattan-turning-against-congestion-pricing/.

Charles-Guzman, Kizzy M. 2012. "Air Pollution Control Strategies in New York City: A Case Study of the Role of Environmental Monitoring, Data Analysis, and Stakeholder Networks in Comprehensive Government Policy Development." Master's Thesis, Ann Arbor: University of Michigan. https://deepblue.lib.umich.edu/bitstream/handle/2027.42/94532/Kizzy%20Charles-Guzman%20practicum%20120312.pdf?sequence=1&isAllowed=y.

Chester, Mikhail V., Josh Sperling, Eleanor Stokes, Braden Allenby, Kara Kockelman, Christopher Kennedy, Lawrence A. Baker, James Keirstead, and Chris T. Hendrickson. 2014. "Positioning Infrastructure and Technologies for Low-Carbon Urbanization." *Earth's Future* 2 (10): 2014EF000253. https://doi.org/10.1002/2014EF000253.

Chronopoulos, Themis. 2012. "Congestion Pricing: The Political Viability of a Neoliberal Spatial Mobility Proposal in London, Stockholm, and New York City. *Urban Research & Practice* 5 (2). http://www.tandfonline.com/doi/abs/10.1080/17535069.2012.691617?journalCode=rurp20#.VvlLhvkrI2w.

City of Los Angeles. 2007. *Green LA: An Action Plan to Lead the Nation in Fighting Global Warming*. Los Angeles, CA: Office of the Mayor.

———. 2008. "ClimateLA: Municipal Program Implementing the *GreenLA* Climate Action Plan." Los Angeles, CA.

———. 2015. "*PLAn*: Transforming Los Angeles." Los Angeles, CA: Office of the Mayor.

———. 2016. "Los Angeles Climate Action Report: Updated 1990 Baseline and 2013 Emissions Inventory Summary." Los Angeles, CA.

City of New York. 2007a. "Comprehensive Annual Financial Report of the Comptroller." New York, NY.

———. 2007b. "*PlaNYC*: A Greener, Greater New York." New York, NY: Mayor's Office of Long-Term Planning and Sustainability.

———. 2013. "*PlaNYC* 2013 Progress Report: A Greener, Greater New York." www.planyc_progress_report_2013.

———. 2014. "*PlaNYC* 2014 Progress Report: A Greener, Greater New York. A Stronger, More Resilient New York." http://140422_plaNYC_report_FINAL_Web.

———. 2016. *New York City's Roadmap to 80×50.* New York: Office of the Mayor.
City of Toronto. n.d. "Carbon Credit Policy." http://www1.toronto.ca/wps/portal/contentonly?vgnextoid=e8c4fbfa98491410VgnVCM10000071d60f89RCRD&vgnextchannel=a201fbfa98491410VgnVCM10000071d60f89RCRD.
———. 1991. "The Changing Atmosphere: Strategies for Reducing CO2 Emissions." Toronto: City of Toronto.
———. 2007a. *Change is in the Air.* Toronto: Toronto Environment Office and Toronto Energy Efficiency Office.
———. 2007b. "Greenhouse Gas and Air Pollutants in the City of Toronto." Toronto: Toronto Atmospheric Fund and Toronto Environment Office.
———. 2015. "Toronto's 2013 Greenhouse Gas Emissions Inventory." Toronto: Chief Corporate Officer. http://www.toronto.ca/legdocs/mmis/2016/pe/bgrd/backgroundfile-87697.pdf.
———. 2016a. *TransformTO.* http://www1.toronto.ca/wps/portal/contentonly?vgnextoid=ba07f60f4adaf410VgnVCM10000071d60f89RCRD.
———. 2016b. "*TransformTO*: Climate Action for a Healthy, Equitable, and Prosperous Toronto - Report #1." Staff Report PY2.1. Toronto: Energy and Environment Division.
Clarke, Susan E. 2016. "Local Place-Based Collaborative Governance Comparing State-Centric and Society-Centered Models." *Urban Affairs Review*, 1078087416637126.
Clarke, Susan E., and Gary L. Gaile. 1998. *The Work of Cities.* Minneapolis: University of Minnesota.
Clean Air Partnership. 2007. "Cities Preparing for Climate Change: A Study of Six Urban Regions." Toronto.
Climate Group. 2005. "Low Carbon Leaders: Cities Oct. 2005." London: C20 Large City Network.
Climate Institute. 2015. "Are Countries on Track to Meet Emissions Targets?" Sydney, Australia. http://www.climateinstitute.org.au/verve/_resources/TCI_On_track_factsheet.pdf.
Cohen, S., and A. Miller. 2012. "Climate Change 2011: A Status Report on US Policy." *Bulletin of the Atomic Scientists* 68 (1): 39–49. https://doi.org/10.1177/0096340211433007.
The Conservation Law Foundation. 2010. "Case Study: 2008 Los Angeles County Ballot Measure to Pass a Transportation Sales Tax (Measure R)." Chicago. http://www.northeastern.edu/dukakiscenter/wp-content/uploads/Lessons-Learned-Brief-2010-LA-Measure-R-Campaign.pdf.
Dahlberg, Peter, Emma O'Brien, and Nick Zoller. 2007. "Bloomberg's Folly? Congestion Pricing in New York City." http://nexus.umn.edu/Courses/Cases/CE5212/F2007/CS8/CS8-report.pdf.
Davenport, Coral. 2015. "U.S. and Chinese Climate Change Negotiators to Meet in Los Angeles." *New York Times*, September 15 edition, sec. U.S. http://mobile.nytimes.com/2015/09/16/us/us-and-chinese-climate-change-negotiators-to-meet-in-los-angeles.html.
DeShazo, J. R., Alex Turek, and Michael Samulon. 2014. "Efficiently Energizing Job Creation in Los Angeles." Los Angeles: University of California Los Angeles Luskin Center for Innovation. http://innovation.luskin.ucla.edu/sites/default/files/UCLA-LADWP%20EE%20Jobs%20Study_0.pdf.
Dhanak, Shawn, and Shane Levy. 2016. "Los Angeles Takes Major Step Toward 100% Clean Energy." *Sierra Club*, September 16. http://content.sierraclub.org/press-releases/2016/09/los-angeles-takes-major-step-toward-100-clean-energy.

Dolan, Jack. 2016. "Advocate of DWP Reform Might Have the Most to Gain." *Los Angeles Times*, November 6. http://www.latimes.com/local/la-me-adv-dwp-ballot-measure-20161031-story.html.

Dunn, Christopher. 2008. "Canada's 'Open Federalism': Past, Present, and Future." In *The Federal Nation: Perspectives on American Federalism*, by Iwan W. Morgan and Philip J. Davies, 39–60. New York: Palgrave Macmillan.

Emerson, Kirk, and Tina Nabatchi. 2015. *Collaborative Governance Regimes*. Washington, DC: Georgetown University Press.

Environics Institute for Survey Research. 2015. "Focus Canada 2015: Canadian Public Opinion about Climate Change." David Suzuki Foundation and The Environics Institute. www.davidsuzuki.org/publications/downloads/2015/Environics Final Report 2015.pdf.

FCM and ICLEI. 2010. "Demonstrating Results: Municipal Initiatives to Reduce Greenhouse Gases." Partners for Climate Protection: National Measures Report.

Federation of Canadian Municipalities. 2009. "Act Locally: The Municipal Role in Fighting Climate Change." Ottawa. https://www.fcm.ca/Documents/report/Act_Locally_The_Municipal_Role_in_Fighting_Climat_Change_EN.pdf.

———. 2015a. "20 Years of Taking Action." Ottawa. http://www.fcm.ca/home/programs/partners-for-climate-protection/20-years-of-taking-action.htm.

———. 2015b. "Partners for Climate Protection: National Measures Report 2015." Ottawa: The Federation of Canadian Municipalities.

———. 2017. "Partners for Climate Protection." http://www.fcm.ca/home/programs/partners-for-climate-protection.htm.

Feiock, Richard C., ed. 2004. *Metropolitan Governance: Conflict, Competition, and Cooperation*. Washington, DC: Georgetown University Press.

Florida, Richard. 2002. *The Rise of the Creative Class: And How It's Transforming Work, Leisure, Community and Everyday Life*. New York: Basic Books.

Fong, Wee Kean, Mary Sotos, Michael Doust, Sean Schultz, Ana Marques, and Chang Deng-Beck. 2015. "Global Protocol for Community-Scale Greenhouse Gas Emission Inventorie: An Accounting and Reporting Standard for Cities." World Resources Institute, C40 Cities, and ICLEI Local Governments for Sustainability.

Frug, Gerald E., and David J. Barron. 2008. *City Bound: How States Stifle Urban Innovation*. Ithaca, NY: Cornell University Press.

Gerdes, Justin. 2013. "Copenhagen's Ambitious Push to Be Carbon Neutral by 2025." *Yale Environment 360*, April 11. http://e360.yale.edu/features/copenhagens_ambitious_push_to_be_carbon_neutral_by_2025.

Ghertner, D. Asher. 2012. "Nuisance Talk and the Propriety of Property: Middle Class Discourses of a Slum-Free Delhi." *Antipode* 44 (4): 1161–1187. https://doi.org/10.1111/j.1467-8330.2011.00956.x.

Giambusso, David. 2016. "De Blasio's Energy Goals Earn Praise, but Challenges Remain." *Politico*, February 19. http://www.politico.com/states/new-york/city-hall/story/2016/02/de-blasios-energy-goals-earn-praise-but-challenges-remain-031410.

Giest, Sarah. 2017. "Big Data Analytics for Mitigating Carbon Emissions in Smart Cities: Opportunities and Challenges." *European Planning Studies* 25 (6): 941–957.

Glaeser, Edward. 2011. *Triumph of the City*. New York: Penguin Press.

Global Covenant of Mayors. n.d. "Fact Sheet." https://www.bbhub.io/mayors/sites/14/2016/06/Global-Covenant-of-Mayors-for-Climate-Energy-Fact-Sheet-FINAL.pdf.

Good, Kristin. 2009. *Municipalities and Multiculturalism: The Politics of Immigration in Toronto and Vancouver*. Toronto: University of Toronto Press.

Goodling, Erin, Jamaal Green, and Nathan McClintock. 2015. "Uneven Development of the Sustainable City: Shifting Capital in Portland, Oregon." *Urban Geography* 36 (4): 504–527.
Gordon, David J. 2015. "From Global Cities to Global Governors: Power, Politics and the Convergence of Urban Climate Governance." Unpublished doctoral dissertation, University of Toronto.
Gore, Christopher D. 2010. "The Limits and Opportunities of Networks: Municipalities and Canadian Climate Change Policy." *Review of Policy Research* 27 (1): 27–46. https://doi.org/10.1111/j.1541-1338.2009.00425.x.
Gore, Christopher D., and Pamela J. Robinson. 2009. "Local Government Responses to Climate Change: Our Last, Best Hope?" In *Changing Climates in North American Politics: Institutions, Policymaking and Multilevel Governance*, by Henrik Selin and Stacy D. VanDeveer, 138–158. Cambridge, MA: MIT Press.
Gore, Christopher, Pamela Robinson, and Richard Stren. 2009. "Governance and Climate Change: Assessing and Learning from Canadian Cities." Marseille, France: Fifth Urban Research Symposium Cities and Climate Change: Responding to an Urgent Agenda.
Gössling, Stefan. 2013. "Urban Transport Transitions: Copenhagen, City of Cyclists." *Journal of Transport Geography* 33 (December): 196–206. https://doi.org/10.1016/j.jtrangeo.2013.10.013.
Gottlieb, Robert, Mark Vallianatos, Regina M. Freer, and Peter Dreier. 2006. *The Next Los Angeles: The Struggle for a Livable City*. Second edition. Los Angeles and Berkeley: University of California Press.
Greasley, Stephen, and Gerry Stoker. 2009. "Urban Political Leadership." In *Theories of Urban Politics*, edited by Jonathan S. Davies and David L. Imbroscio, 125–136. Second edition. London: Sage.
Gronewold, Nathanial. 2010. "NYC Begins Hard, Long Slog to Energy Efficiency." *New York Times*, April 5, sec. Business/Energy & Environment. http://www.nytimes.com/gwire/2010/04/05/05greenwire-nyc-begins-hard-long-slog-to-energy-efficiency-78815.html.
Gross, Jill Simone. 2017. "Hybridization and Urban Governance: Malleability, Modality, or Mind-Set?" *Urban Affairs Review* 53 (3): 559–577.
Hackworth, Jason. 2015. "Rightsizing as Spatial Austerity in the American Rust Belt." *Environment and Planning A* 47 (4): 766–782.
———. 2016. "Demolition as Urban Policy in the American Rust Belt." *Environment and Planning A* 48 (11): 2201–2222.
Hackworth, Jason, and Abigail Moriah. 2006. "Neoliberalism, Contingency and Urban Policy: The Case of Social Housing in Ontario." *International Journal of Urban and Regional Research* 30 (3): 510–527. https://doi.org/10.1111/j.1468-2427.2006.00675.x.
Haldi, Frédéric, and Darren Robinson. 2011. "The Impact of Occupants' Behaviour on Building Energy Demand." *Journal of Building Performance Simulation* 4 (4): 323–338. https://doi.org/10.1080/19401493.2011.558213.
Halper, Evan. 2017. "A California-Led Alliance of Cities and States Vows to Keep the Paris Climate Accord in Tact." *New York Times*, June 2, sec. Politics. http://www.latimes.com/politics/la-na-pol-paris-states-20170602-story.html.
Hambleton, Robin, and Jill Simone Gross. 2007. *Governing Cities in a Global Era*. Basingstoke: Palgrave Macmillan.
Harris, Sara E., and Sarah L. Burch. 2014. *Understanding Climate Change: Science, Policy, and Practice*. Toronto: University of Toronto Press.

Harvey, L. D. Danny. 1993. "Tackling Urban CO_2 Emissions in Toronto." *Environment: Science and Policy for Sustainable Development* 35 (7): 16–44. https://doi.org/10.1080/00139157.1993.9929991.

Hawkins, Christopher V., Rachel M. Krause, Richard C. Feiock, and Cali Curley. 2015. "Making Meaningful Commitments: Accounting for Variation in Cities' Investments of Staff and Fiscal Resources to Sustainability." *Urban Studies*, April, 0042098015580898. https://doi.org/10.1177/0042098015580898.

Hawkins, Christopher V., and XiaoHu Wang. 2012. "Sustainable Development Governance Citizen Participation and Support Networks in Local Sustainability Initiatives." *Public Works Management & Policy* 17 (1): 7–29. https://doi.org/10.1177/1087724X11429045.

Hayward, Clarissa Rile, and Todd Swanstrom. 2011. *Justice and the American Metropolis*. Minneapolis, MN: University of Minnesota Press.

Healey, Patsy, S. Cameron, S. Davoudi, S. Graham, and A. Madani-Pour, eds. 1995. *Managing Cities: The New Urban Context*. Chichester: John Wiley & Son.

Heitzig, Jobst, Kai Lessmann, and Yong Zou. 2011. "Self-Enforcing Strategies to Deter Free-Riding in the Climate Change Mitigation Game and Other Repeated Public Good Games." *Proceedings of the National Academy of Sciences* 108 (38): 15739–15744.

Hempel, Lamont C. 2003. "Climate Policy on the Installment Plan." In *Environmental Policy: New Directions for the Twenty-First Century*, by Norman J. Vig and Michael E. Kraft, 299–323. Washington, DC: CQ Press.

Hjern, Benny. 1982. "Implementation Research—The Link Gone Missing." *Journal of Public Policy* 2: 301–308.

Hjerpe, Mattias, and Naghmeh Nasiritousi. 2015. "Views on Alternative Forums for Effectively Tackling Climate Change." *Nature Climate Change* advance online publication (June). https://doi.org/10.1038/nlcimate2684.

Hodson, Michael, and Simon Marvin. 2010. "Can Cities Shape Socio-Technical Transitions and How Would We Know If They Were?" *Research Policy* 39 (4): 477–485.

Hoffmann, Matthew J. 2011. *Climate Governance at the Crossroads: Experimenting with a Global Response*. New York: Oxford University Press.

Homsy, George, and Mildred E. Warner. 2015. "Cities and Sustainability: Polycentric Action and Multilevel Governance." *Urban Affairs Review* 51 (1): 46–73.

Horak, Martin. 2013. "State Rescaling in Practice: Urban Governance Reform in Toronto." *Urban Research & Practice* 6 (3): 311–328.

Hu, Winnie. 2018. "Congestion Pricing Falters in New York, Again." *New York Times*, March 31. https://www.nytimes.com/2018/03/31/nyregion/congestion-pricing-new-york.html.

Hughes, Sara. 2017. "The Politics of Urban Climate Change Policy: Toward a Research Agenda." *Urban Affairs Review* 53 (2): 362–380. https://doi.org/10.1177/1078087416649756.

Hughes, Sara, Stephanie Pincetl, and Christopher G. Boone. 2013. "Understanding Major Urban Transitions: Drivers and Dynamics in the City of Los Angeles." *Cities: The International Journal of Urban Policy and Planning* 32: 51–59.

Hughes, Sara, Laura Tozer, and Sarah Giest. Forthcoming. "The Politics of Data-Driven Urban Climate Change Mitigation." In *Urban Climate Politics: Agency and Empowerment*, by Jeroen van der Heijden, Harriet Bulkeley, and Chiara Certoma. Cambridge University Press.

Hunt, Alistair, and Paul Watkiss. 2011. "Climate Change Impacts and Adaptation in Cities: A Review of the Literature." *Climatic Change* 104 (1): 13–49.

ICLEI. 2010. "The Process Behind *PlaNYC*: How the City of New York Developed Its Comprehensive Long-Term Sustainability Plan." ICLEI-Local Governments for Sustainability USA. http://s-media.nyc.gov/agencies/planyc2030/pdf/iclei_planyc_case_study_201004.pdf.

———. 2015. "Measuring Up 2015: How Cities Are Accelerating Progress Toward National Climate Goals." Oakland, CA: ICLEI-Local Governments for Sustainability USA.

Imbroscio, David L. 1997. *Reconstructing City Politics: Alternative Economic Development and Urban Regimes*. Thousand Oaks, CA: Sage.

Imbroscio, David L., Thad Williamson, and Gar Alperovitz. 2003. "Local Policy Responses to Globalization: Place-Based Ownership Models of Economic Enterprise." *Policy Studies Journal* 31 (1): 31–52. https://doi.org/10.1111/1541-0072.00002.

Institute for Building Efficiency. 2010. "The Big Apple Takes Bold Steps Toward Energy Efficiency in Buildings." November 2010. http://www.institutebe.com/energy-policy/new-york-building-efficiency.aspx.

IPCC. 2011. "Renewable Energy Sources and Climate Change Mitigation." Cambridge and New York: Cambridge University Press.

———. 2013. "Summary for Policymakers." In *Climate Change 2013: The Physical Science Basis. Contribution of Working Group I to the Fifth Assessment Report of the Intergovernmental Panel on Climate Change*. Cambridge and New York: Cambridge University Press.

———. 2014a. "Climate Change 2014: Impacts, Adaptation, and Vulnerability. Contribution of Working Group II to the Fifth Assessment Report of the Intergovernmental Panel on Climate Change." Cambridge and New York: Cambridge University Press.

———. 2014b. "Contribution of Working Group III to the Fifth Assessment Report of the Intergovernmental Panel on Climate Change." Cambridge and New York: Cambridge University Press. http://mitigation2014.org/report/publication/.

Jacobs, Alan. 2011. *Governing for the Long Term: Democracy and the Politics of Investment*. New York: Cambridge University Press.

Janda, Kathryn B. 2011. "Buildings Don't Use Energy: People Do." *Architectural Science Review* 54 (1): 15–22. https://doi.org/10.3763/asre.2009.0050.

Jessop, Bob. 1990. *State Theory: Putting the Capitalist State in Its Place*. University Park: Pennsylvania University Press.

Jessop, Bob, Jamie Peck, and Adam Tickell. 1999. "Retooling the Machine Economic Crisis, State Restructuring, and Urban Politics." In *The Urban Growth Machine: Critical Perspectives, Two Decades Later*, 141–162. New York: State University of New York Press.

Jessop, Bob, and Ngai-Ling Sum. 2000. "An Entrepreneurial City in Action: Hong Kong's Emerging Strategies in and for (Inter)Urban Competition." *Urban Studies* 37 (12): 2287–2313. https://doi.org/10.1080/00420980020002814.

Jones, Bryan D. 1983. *Governing Urban America*. Boston: Little, Brown.

Jones, C. and D. M. Kammen. 2014. "Spatial distribution of U.S. household carbon footprints reveals suburbanization undermines greenhouse gas benefits of urban population density." *Environmental Science and Technology* 48 (2): 895–902.

Judd, Dennis R., and Todd Swanstrom. 2012. *City Politics*. 8th ed. Glenview, IL: Pearson Education.

Katz, Bruce, and Jennifer Bradley. 2013. *Metropolitan Revolution: Los Angeles*. Washington, DC: Brookings Institution.

Katz, Bruce, and Jeremy Nowak. 2017. *The New Localism*. Washington, DC: Brookings Institution Press.

Keil, Roger. 2000. "Governance Restructuring in Los Angeles and Toronto: Amalgamation or Secession?" *International Journal of Urban and Regional Research* 24 (4): 758–781.

Kennedy, Christopher A., Iain Stewart, Angelo Facchini, Igor Cersosimo, Renata Mele, Bin Chen, Mariko Uda, et al. 2015. "Energy and Material Flows of Megacities." *Proceedings of the National Academy of Sciences*, April, 201504315. https://doi.org/10.1073/pnas.1504315112.

Kern, Kristine, and Harriet Bulkeley. 2009. "Cities, Europeanization and Multi-Level Governance: Governing Climate Change through Transnational Municipal Networks." *JCMS: Journal of Common Market Studies* 47 (2): 309–332.

Kettl, Donald F. 2000. "The Transformation of Governance: Globalization, Devolution, and the Role of Government." *Public Administration Review* 60 (6): 488–497. https://doi.org/10.1111/0033-3352.00112.

Kinman, Michelle. 2016. "Los Angeles to Chart Path to 100% Renewables." *Environment California*, September 16. http://www.environmentcalifornia.org/news/cae/los-angeles-chart-path-100-renewable-energy.

Kjaer, Anne Mette. 2009. "Governance and the Urban Bureaucracy." In *Theories of Urban Politics*, edited by Jonathan S. Davies and David L. Imbroscio, 137–152. Second edition. London: Sage Publications.

Klyza, Christopher McGrory, and David Sousa. 2008. *American Environmental Policy, 1990-2006*. Cambridge, MA: MIT Press.

Koontz, Tomas M., Toddi A. Steelman, JoAnn Carmin, Katrina Smith Korfmacher, Cassandra Mosley, and Craig W. Thomas. 2004. *Collaborative Environmental Management: What Roles for Government?* Washington, DC: Resources for the Future.

Koppenjan, Joop, and Erik-Hans Klijn. 2004. *Managing Uncertainties in Networks: A Network Approach to Problem Solving and Decision Making*. London and New York: Routledge.

Koski, Chris. 2010. "Greening America's Skylines: The Diffusion of Low-Salience Policies." *Policy Studies Journal* 38 (1): 93–117. https://doi.org/10.1111/j.1541-0072.2009.00346.x.

Koven, S., and T. Lyons. 2010. "Economic Development: Strategies for State and Local Practice." Washington, DC: International City/County Management Association.

Krause, Rachel M. 2011a. "Policy Innovation, Intergovernmental Relations, and the Adoption of Climate Protection Initiatives by US Cities." *Journal of Urban Affairs* 33 (1): 45–60.

———. 2011b. "Symbolic or Substantive Policy? Measuring the Extent of Local Commitment to Climate Protection." *Environment and Planning C: Government and Policy* 29 (1): 46–62.

———. 2012. "Political Decision-Making and the Local Provision of Public Goods: The Case of Municipal Climate Protection in the US." *Urban Studies* 49 (11): 2399–2417. https://doi.org/10.1177/0042098011427183.

Krause, Rachel M., Hongtao Yi, and Richard C. Feiock. 2015. "Applying Policy Termination Theory to the Abandonment of Climate Protection Initiatives by U.S. Local Governments." *Policy Studies Journal*, June, n/a-n/a. https://doi.org/10.1111/psj.12117.

Lambright, W. Henry, Stanley A. Chjangnon, and L. D. Danny Harvey. 1996. "Urban Reactions to the Global Warming Issue: Agenda Setting in Toronto and Chicago." *Climatic Change* 34 (3–4): 463–478. https://doi.org/10.1007/BF00139302.

Laurian, Lucie, Mark Walker, and Jan Crawford. 2016. "Implementing Environmental Sustainability in Local Government: The Impacts of Framing, Agency Culture,

and Structure in US Cities and Counties." *International Journal of Public Administration*, 1–15.
Leach, Andrew. 2011. "The Nuts and Bolts of Kyoto Withdrawal." *The Globe and Mail*, December 7. https://www.theglobeandmail.com/report-on-business/economy/economy-lab/the-nuts-and-bolts-of-kyoto-withdrawal/article619868/.
Lee, Taedong. 2012. "Global Cities and Transnational Climate Change Networks." *Global Environmental Politics* 13 (1): 108–127. https://doi.org/10.1162/GLEP_a_00156.
Lee, Taedong, and Chris Koski. 2012. "Building Green: Local Political Leadership Addressing Climate Change." *Review of Policy Research* 29 (5): 605–624.
Leiserowitz, Anthony, E. Maibach, C. Roser-Renouf, and S Rosenthal. 2016. "Politics and Global Warming, Spring 2016." New Haven, CT: Yale Program on Climate Change Communication: Yale University and George Mason University.
Los Angeles Office of the Mayor. 2014. "Mayor Garcetti Highlights LADWP's New Energy Efficiency Goal, Which Leads Nation," https://www.lamayor.org/mayor-garcetti-highlights-ladwp%E2%80%99s-new-15-energy-efficiency-goal-which-leads-nation. Accessed January 23, 2019.
Los Angeles Times Editorial Board. 2016. "Proposed DWP Rate Hike Isn't High Enough." *Los Angeles Times*, March 1. http://www.latimes.com/opinion/editorials/la-ed-0301-dwp-rate-hikes-20160216-story.html.
Lubell, Mark, Richard Feiock, and Susan Handy. 2009. "City Adoption of Environmentally Sustainable Policies in California's Central Valley." *Journal of the American Planning Association* 75 (3): 293–308. https://doi.org/10.1080/01944360902952295.
Lutsey, Nicholas, and Daniel Sperling. 2008. "America's Bottom-up Climate Change Mitigation Policy." *Energy Policy* 36 (2): 673–685. https://doi.org/10.1016/j.enpol.2007.10.018.
Marcotullio, Peter J., Sara Hughes, Andrea Sarzynski, Stephanie Pincetl, Landy Sanchez Peña, Patricia Romero-Lankao, Daniel Runfola, and Karen C. Seto. 2014. "Urbanization and the Carbon Cycle: Contributions from Social Science." *Earth's Future* 2 (10): 2014EF000257. https://doi.org/10.1002/2014EF000257.
Mathur, Vivek Narain, Andrew D. F. Price, and Simon Austin. 2008. "Conceptualizing Stakeholder Engagement in the Context of Sustainability and Its Assessment." *Construction Management and Economics* 26 (6): 601–609. https://doi.org/10.1080/01446190802061233.
Matsuoka, Martha M., and Robert Gottlieb. 2013. "Environmental and Social Justice Movements and Policy Change in Los Angeles: Is an Inside-Out Game Possible?" In *New York and Los Angeles: The Uncertain Future*, by David Halle and Andrew A. Beveridge, 445–468. Oxford and New York: Oxford University Press.
Mattern, Sara. 2013. "Municipal Energy Benchmarking Legislation for Commercial Buildings: You Can't Manage What You Don't Measure." *Boston College Environmental Affairs Law Review* 40 (2): 487.
Mayor's Office of Long-Term Planning and Sustainability. 2014. "One City, Built to Last: Transforming New York City's Buildings for A Low-Carbon Future." http://www.nyc.gov/html/builttolast/assets/downloads/pdf/OneCity.pdf.
Mazmanian, Daniel A., and Michael E. Kraft. 2009. *Toward Sustainable Communities: Transition and Transformations in Environmental Policy*. Second edition. Cambridge, MA: MIT Press.
Mazmanian, Daniel A., and Paul A. Sabatier. 1981. *Effective Policy Implementation*. Lexington, MA: Heath Publishers.
McCright, Aaron M., and Riley E. Dunlap. 2011. "The Politicization of Climate Change and Polarization in the American Public's Views of Global Warming, 2001–2010."

Sociological Quarterly 52 (2): 155–194. https://doi.org/10.1111/j.1533-8525.2011.01198.x.

McGuirk, Pauline M. 2003. "Producing the Capacity to Govern in Global Sydney: A Multiscaled Account." *Journal of Urban Affairs* 25 (2): 201–223.

McKendry, Corina. 2016. "Cities and the Challenge of Multiscalar Climate Justice: Climate Governance and Social Equity in Chicago, Birmingham, and Vancouver." *Local Environment* 21 (11): 1354–1371.

McNeill, J. R. 2000. *Something New Under the Sun: An Environmental History of the Twentieth-Century World*. New York: W. W. Norton.

Meadowcroft, James. 2007. "Who Is in Charge Here? Governance for Sustainable Development in a Complex World*." *Journal of Environmental Policy & Planning* 9 (3–4): 299–314.

Melillo, Jerry M., Terese C. Richmond, and Gary Yohe. 2014. "Highlights of Climate Change Impacts in the United States: The Third National Climate Assessment." U.S. Global Change Research Program.

Mitchell, Ronald B. 2008. "Evaluating the Performance of Environmental Institutions: What to evaluate and how to evaluate it?" In *Institutions and Environmental Change: Principal Findings, Applications, and Research Frontiers*, edited by Oran R. Young, Leslie A. King, and Heike Schroeder, 79–114. Cambridge, MA: MIT Press.

Molotch, Harvey. 1976. "The City as a Growth Machine: Toward a Political Economy of Place." *American Journal of Sociology*, 309–332.

Monstadt, Jochen. 2009. "Conceptualizing the Political Ecology of Urban Infrastructures: Insights from Technology and Urban Studies." *Environment and Planning A* 41: 1924–1942.

Mossberger, Karen. 2009. "Urban Regime Analysis." In *Theories of Urban Politics*, edited by Jonathan S. Davies and David L. Imbroscio, 40–54. Second edition. London: Sage Publications.

Mullin, Megan, Gillian Peele, and Bruce E. Cain. 2004. "City Caesars? Institutional Structure and Mayoral Success in Three California Cities," *Urban Affairs Review* 40 (1): 19–43.

Nelson, Laura J. 2016. "Los Angeles Area Can Claim the Worst Traffic in America. Again." *Los Angeles Times*, March 15. http://www.latimes.com/local/lanow/la-me-ln-la-worst-traffic-20160314-story.html.

Neumann, Tracy. 2016. *Remaking the Rust Belt: The Postindustrial Transformation of North America*. Philadelphia: University of Pennsylvania Press.

Newton, Damien. 2011. "Goodbye, 30/10. Hello, Fast Forward America." *StreetsBlog LA*, February 23. https://la.streetsblog.org/2011/02/23/field-hearing-report-villaraigosa-rebrands-l-a-s-transit-funding-plan-for-as-one-for-all-america/.

New York City Energy Policy Task Force. 2004. "New York City Energy Policy: An Electricity Resource Roadmap." New York, NY. http://www.nyc.gov/html/om/pdf/energy_task_force.pdf.

New York Department of Transportation. 2008. "Sustainable Streets: Strategic Plan for the New York City Department of Transportation 2008 and Beyond." New York, NY. http://www.nyc.gov/html/dot/downloads/pdf/stratplan_compplan.pdf.

Nicholls, Walter J. 2005. "Power and Governance: Metropolitan Governance in France." *Urban Studies* 42 (4): 783–800.

Office of the Mayor. 2015a. "De Blasio Administration Moves to Power 100 Percent of City Government From Renewable Sources of Energy." http://www1.nyc.gov/office-of-the-mayor/news/478-15/de-blasio-administration-moves-power-100-percent-city-government-renewable-sources-of.

———. 2015b. "Solar Installations More Than Doubled Since Beginning of de Blasio Administration, Mayor Announces During Climate Week." http://www1.nyc.gov/office-of-the-mayor/news/631-15/solar-installations-more-doubled-since-beginning-de-blasio-administration-mayor-announces.

O'Neil, Lauren. 2018. "Every Ontario Ministry Now Banned from Talking about Climate Change." *BlogTO*, August 25. https://www.blogto.com/city/2018/08/doug-ford-climate-change/.

Ontario Ministry of Environment, Conservation, and Parks. 2017. "Ontario Announces Results of First Cap and Trade Program Auction." https://news.ontario.ca/ene/en/2017/04/ontario-announces-results-of-first-cap-and-trade-program-auction.html.

Osborne, D., and T. Gaebler. 1991. *Reinventing Government*. Reading, MA: Addison-Wesley.

O'Toole, Laurence J. 2000. "Research on Policy Implementation: Assessment and Prospects." *Journal of Public Administration Research and Theory* 10 (2): 263–288.

Peters, B. Guy. 1997. "Shouldn't Row, Can't Steer: What's a Government to Do?" *Public Policy and Administration* 12 (2): 51–61.

Peterson, Jacqueline. 2018. "Multilevel Governance and Innovations in the Financing of Urban Climate Change Strategies." In *Climate Change in Cities: Innovations in Multi-Level Governanc*, edited by Sara Hughes, Eric K. Chu, and Susan G. Mason Cham. Switzerland: Springer.

Peterson, Paul E. 1981. *City Limits*. Chicago: Chicago University Press.

Pierre, Jon. 2011. *The Politics of Urban Governance*. New York: Palgrave.

———. 2014. "Can Urban Regimes Travel in Time and Space? Urban Regime Theory, Urban Governance Theory, and Comparative Urban Politics." *Urban Affairs Review* 50 (6): 864–889.

Pierre, Jon, and B. Guy Peters. 2012. "Urban Governance." In *The Oxford Handbook of Urban Politics*, edited by Karen Mossberger, Susan E. Clarke, and Peter John, 71–86. New York: Oxford University Press.

Pitt, Damian Rogero. 2010. "Harnessing Community Energy: The Keys to Climate Mitigation Policy Adoption in US Municipalities." *Local Environment* 15 (8): 717–729. https://doi.org/10.1080/13549839.2010.509388.

Pitt, Damian, and John Randolph. 2009. "Identifying Obstacles to Community Climate Protection Planning." *Environment and Planning C: Government and Policy* 27 (5): 841–857. https://doi.org/10.1068/c0871.

Portney, Kent E. 2013. *Taking Sustainable Cities Seriously: Economic Development, the Environment, and Quality of Life in American Cities*. Second Edition. Cambridge, MA: MIT Press. http://www.tandfonline.com/doi/pdf/10.1080/08111114032000113680.

Portney, Kent E., and Jeffrey M. Berry. 2010. "Participation and the Pursuit of Sustainability in U.S. Cities." *Urban Affairs Review* 46 (1): 119–39. https://doi.org/10.1177/1078087410366122.

Port of Los Angeles. 2016. "Facts and Figures." Los Angeles, CA. https://www.portoflosangeles.org/pdf/POLA_Facts_and_Figures_Card.pdf.

Qin, Yuwei, Shannon Rivers, Jennifer Roecks, and Francisca Santana. 2014. "Local Government Climate Action Plan Implementation in California." Santa Barbara, CA: University of California Bren School of Environmental Science & Management.

Rabe, Barry G. 2016. "The Durability of Carbon Cap-and-Trade Policy." *Governance* 29 (1): 103–119. https://doi.org/10.1111/gove.12151.

Ramaswami, Anu, Tim Hillman, Bruce Janson, Mark Reiner, and Gregg Thomas. 2008. "A Demand-Centered, Hybrid Life-Cycle Methodology for City-Scale Greenhouse

Gas Inventories." *Environmental Science & Technology* 42 (17): 6455–6461. https://doi.org/10.1021/es702992q.

Reames, Tony Gerard. 2016. "Targeting Energy Justice: Exploring Spatial, Racial/Ethnic and Socioeconomic Disparities in Urban Residential Heating Energy Efficiency." *Energy Policy* 97: 549–558.

Reams, Margaret A., Kelsey W. Clinton, and Nina S. N. Lam. 2012. "Achievement of Climate Planning Objectives among U.S. Member Cities of the International Council for Local Environmental Initiatives (ICLEI)." *Low Carbon Economy* 3 (4): 137–143. https://doi.org/10.4236/lce.2012.34018.

Rich, Michael J., and Robert P. Stoker. 2015. *Collaborative Governance for Urban Revitalization: Lessons from Empowerment Zones*. Ithaca, NY: Cornell University Press.

Rittel, Horst W. J., and Melvin M. Webber. 1973. "Dilemmas in a General Theory of Planning." *Policy Sciences* 4 (2): 155–169.

Robinson, Pamela J., and Christopher D. Gore. 2005. "Barriers to Canadian Municipal Response to Climate Change." *Canadian Journal of Urban Research* 14 (1): 102–120.

Rockström, Johan, Owen Gaffney, Joeri Rogelj, Malte Meinshausen, Nebojsa Nakicenovic, and Hans Joachim Schellnhuber. 2017. "A Roadmap for Rapid Decarbonization." *Science* 355 (6331): 1269–1271. https://doi.org/10.1126/science.aah3443.

Romero-Lankao, Patricia. 2007. "Are We Missing the Point?" *Environment and Urbanization* 19 (1): 159–175. https://doi.org/10.1177/0956247807076915.

Romero-Lankao, Patricia, Kevin R. Gurney, Karen C. Seto, Mikhail Chester, Riley M. Duren, Sara Hughes, Lucy R. Hutyra, et al. 2014. "A Critical Knowledge Pathway to Low-Carbon, Sustainable Futures: Integrated Understanding of Urbanization, Urban Areas, and Carbon." *Earth's Future* 2 (10): 2014EF000258. https://doi.org/10.1002/2014EF000258.

Romero-Lankao, Patricia, Sara Hughes, Angelica Rosas-Huerta, Roxana Borquez, and David Gnatz. 2013. "Urban Institutional Response Capacity for Climate Change: An Examination of Construction and Pathways in Mexico City and Santiago." *Environment and Planning C: Government and Policy* 31: 785–805.

Rosan, Christina D. 2012. "Can *PlaNYC* Make New York City 'Greener and Greater' for Everyone?: Sustainability Planning and the Promise of Environmental Justice." *Local Environment* 17 (9): 959–976. https://doi.org/10.1080/13549839.2011.627322.

Ryan, Daniel. 2015. "From Commitment to Action: A Literature Review on Climate Policy Implementation at City Level." *Climatic Change*, 1–11.

Sancton, Andrew. 2011. *Canadian Local Government: An Urban Perspective*. Oxford and Cambridge, MA: Oxford University Press.

Sapotichne, Joshua, and Bryan D. Jones. 2011. "Setting City Agendas: Power and Policy Change." In *Oxford Handbook of Urban Politics*, edited by Karen Mossberger, Susan Clarke, and Peter John, 76–106. Oxford, UK: Oxford University Press.

Sarzynski, Andrea. 2015. "Public Participation, Civic Capacity, and Climate Change Adaptation in Cities." *Urban Climate* 14 (1): 52–67. https://doi.org/10.1016/j.uclim.2015.08.002.

Sassen, Saskia. 2000. "The Global City: Strategic Site/New Frontier." *American Studies*, 79–95.

Satterthwaite, David. 2008. "Cities' Contribution to Global Warming: Notes on the Allocation of Greenhouse Gas Emissions." *Environment and Urbanization* 20 (2): 539–549. https://doi.org/10.1177/0956247808096127.

Savitch, H. V., and Paul Kantor. 2002. *Cities in the International Marketplace: The Political Economy of Urban Development in North America and Western Europe.* Princeton, NJ: Princeton University Press.

Schreurs, Miranda A. 2010. "Multi-Level Governance and Global Climate Change in East Asia." *Asian Economic Policy Review* 5 (1): 88–105.

Schwartz, Elizabeth. 2016. "Developing Green Cities: Explaining Variation in Canadian Green Building Policies." *Canadian Journal of Political Science* 49 (4): 621–641. https://www.cambridge.org/core/journals/canadian-journal-of-political-science-revue-canadienne-de-science-politique/article/developing-green-cities-explaining-variation-in-canadian-green-building-policies/398C9E996CD0CD6F2FDE4E8C0D9046D8#.

Schwartz, Sam, Jee Mee Kim, Gerard Soffian, and Annie Weinstock. 2009. "A Comprehensive Transportation Policy for the 21st Century: A Case Study of Congestion Pricing in New York City." *NYU Environmental Law Journal* 17 (March): 580–607.

Scruggs, Gregory. 2015. "With Paris City Hall Declaration, World Mayors Throw Down Gauntlet on Climate." *Cityscope*, December 5. http://citiscope.org/habitatIII/news/2015/12/paris-city-hall-declaration-world-mayors-throw-down-gauntlet-climate.

Selin, Henrik, and Stacy D. VanDeveer. 2007. "Political Science and Prediction: What's Next for U.S. Climate Change Policy?" *Review of Policy Research* 24 (1): 1–27.

Sellers, Jefferey M. 2002a. *Governing from Below: Urban Regions and the Global Economy.* Cambridge: Cambridge University Press.

———. 2002b. "The Nation-State and Urban Governance." *Urban Affairs Review* 37 (5): 611–641. https://doi.org/10.1177/107808740203700501.

Seto, Karen C., S. Dhakal, A. Bigio, H. Blanco, G. C. Delgado, D. Dewar, L. Huang, A. Inaba, A. Kansal, S. Lwasa, J.E. McMahon, D.B. Muller, J. Murakami, H. Nagendra, and A. Ramaswami. 2014. "Human settlements, infrastructure and spatial planning." In *Climate Change 2014: Mitigation of Climate Change. Contribution of Working Group III to the Fifth Assessment Report of the Intergovernmental Panel on Climate Change.* New York: Cambridge University Press, 923–1000.

Shabecoff, Philip. 1988. "Global Warming Has Begun, Expert Tells Senate." *The New York Times*, June 24. http://www.nytimes.com/1988/06/24/us/global-warming-has-begun-expert-tells-senate.html?pagewanted=all.

Sharp, Elaine B. 1997. "Policy Process." In *Handbook of Research on Urban Politics and Policy in the United States*, edited by Ronald K. Vogel. Westport, CT: Greenwood Press.

Sharp, Elaine B., Dorothy M. Daley, and Michael S. Lynch. 2011. "Understanding Local Adoption and Implementation of Climate Change Mitigation Policy." *Urban Affairs Review* 47 (3): 433–457. https://doi.org/10.1177/1078087410392348.

Shipan, Charles R., and Craig Volden. 2006. "Bottom-Up Federalism: The Diffusion of Antismoking Policies from U.S. Cities to States." *American Journal of Political Science* 50 (4): 825–843. https://doi.org/10.1111/j.1540-5907.2006.00218.x.

Skelcher, C., H. Sullivan, and S. Jeffares. 2013. *Hybrid Governance in European Cities.* London: Palgrave.

Slack, Enid, and Andre Cote. 2014. "Is Toronto Fiscally Healthy? A Check-up on the City's Finances." 7. IMFG Perspectives. Toronto: Institute for Municipal Finance and Governance.

Smil, Vaclav. 2008. *Energy in Nature and Society: General Energetics of Complex Systems.* Cambridge: MIT Press.

Solecki, William. 2012. "Urban Environmental Challenges and Climate Change Action in New York City." *Environment and Urbanization* 24 (2): 557–573. https://doi.org/10.1177/0956247812456472.

Statistics Canada. 2011. "Table 1.a: Proportion of Workers Commuting to Work by Car, Truck or van, by Public Transit, on Foot, or by Bicycle, Census Metropolitan Areas, 2011." http://www12.statcan.gc.ca/nhs-enm/2011/as-sa/99-012-x/2011003/tbl/tbl1a-eng.cfm.

Steelman, Toddi A. 2010. *Implementing Innovation: Fostering Enduring Change in Environmental and Natural Resource Governance*. Public Management and Change. Washington, DC: Georgetown University Press.

Steinhauer, Jennifer. 2010. "Los Angeles Electric Rate Linked to Solar Power." *New York Times*, March 10. http://www.nytimes.com/2010/03/11/science/earth/11solar.html.

Stern, Nicholas. 2006. "Stern Review on the Economics of Climate Change." London: H. M. Treasury/Cabinet Office. http://www.hm-treasury.gov.uk/sternreview_index.htm.

Stewart, Iris T., Daniel R. Cayan, and Michael D. Dettinger. 2004. "Changes in Snowmelt Runoff Timing in Western North America under a 'Business as Usual' Climate Change Scenario." *Climatic Change* 62 (1–3): 217–232. https://doi.org/10.1023/B:CLIM.0000013702.22656.e8.

Stoker, Gerry. 1989. "A Regime Framework for Implementation Analysis: Cooperation and Reconciliation of Federalist Perspectives," *Review of Policy Research* 9 (1): 29–249.

———. 1995. "Regime Theory and Urban Politics." In *Theories of Urban Politics*, by David Judge, Gerry Stoker, and Harold Wolman, 54–71. First Edition. London: Sage.

———. 1998a. "Governance as Theory: Five Propositions." *International Social Science Journal* 50 (155): 17–28. https://doi.org/10.1111/1468-2451.00106.

———. 1998b. "Theory and Urban Politics." *International Political Science Review / Revue Internationale de Science Politique* 19 (2): 119–29.

Stone, Clarence N. 1989. *Regime Politics: Governing Atlanta 1946–1988*. Lawrence: University Press of Kansas.

———. 1993. "Urban Regimes and the Capacity to Govern: A Political Economy Approach." *Journal of Urban Affairs* 15 (1): 1–28.

———. 2005. "Looking Back to Look Forward Reflections on Urban Regime Analysis." *Urban Affairs Review* 40 (3): 309–341.

———. 2015. "Reflections on Regime Politics From Governing Coalition to Urban Political Order." *Urban Affairs Review* 51 (1): 101–137.

Sunlight Foundation. 2017. "Celebrating Data and Evidence at the 2017 What Works Cities Summit," March 31. https://sunlightfoundation.com/2017/03/31/celebrating-data-and-evidence-at-the-2017-what-works-cities-summit/.

Svara, James H., Anna Read, and Evalina Moulder. 2011. "Breaking New Ground: Promoting Environmental and Energy Programs in Local Government." Washington, DC: IBM Center for the Business of Government.

Svara, James H., Tanya C. Watt, and Hee Soun Jang. 2013. "How Are U.S. Cities Doing Sustainability? Who Is Getting on the Sustainability Train, and Why?" *Cityscape: A Journal of Policy Development and Research* 15 (1): 9–43.

Tatum, Jenna. 2014. "Reducing NYC's Carbon Emissions One Building at a Time." *GreenBiz*, August 21. https://www.greenbiz.com/blog/2014/08/21/nyc-carbon-challenge-buildings.

Torfing, Jacob, B. Guy Peters, Jon Pierre, and Eva Sørensen. 2012. *Interactive Governance: Advancing the Paradigm*. Oxford University Press. http://www.oxfordscholarship.com/view/10.1093/acprof:oso/9780199596751.001.0001/acprof-9780199596751.

Toronto Atmospheric Fund. 2012. "Direct Investments: Three Energy Efficiency Retrofit Transactions Structured as ESPA Financing." Toronto. https://www.toronto.ca/legdocs/mmis/2012/ta/bgrd/backgroundfile-50172.pdf.

———. 2016. "Robert Cooke Co-Op Case Study: A TowerWise Retrofit Project." Toronto. http://taf.ca/wp-content/uploads/2016/10/Robert-Cooke-Coop-Case-Study.pdf.

Tozer, Laura. 2013. "Community Energy Plans in Canadian Cities: Success and Barriers in Implementation." *Local Environment* 18 (1): 20–35.

UNFCCC. 1992. "United Nations Framework Convention on Climate Change." New York: United Nations.

United Nations. 2015. "Regions, States, and Cities Pivotal for Accelerating Climate Action." Lyon, France. http://www.un.org/sustainabledevelopment/blog/2015/07/regions-states-and-cities-pivotal-for-accelerating-climate-action/.

Unruh, Gregory C. 2000. "Understanding Carbon Lock-In." *Energy Policy* 28 (12): 817–830.

Urban Green Council. 2012. "90 by 50: NYC Can Reduce its Carbon Footprint 90% by 2050." New York, NY: Urban Green Council.

U.S. Census Bureau. 2013. "American Community Survey 2013." Washington, DC: U.S. Bureau of Census.

U.S. Department of Energy. 2015. "New York City Benchmarking and Transparency Policy Impact Evaluation Report." http://energy.gov/sites/prod/files/2015/05/f22/DOE%20New%20York%20City%20Benchmarking%20snd%20Transparency%20Policy%20Impact%20Evaluation . . . . pdf.

U.S. Environmental Protection Agency. 2001. "Partnerships and Progress: EPA State and Local Climate Change Program 2001 Progress Report." Washington, DC.

———. 2016. "U.S. Greenhouse Gas Inventory Report: 1990–2014." https://www3..epa.gov/climatechange/ghgemissions/usinventoryreport.html.

Vande Bunte, Matt. 2015. "Grand Rapids Mayor Headed to Paris for UN Conference." *MLive*, November 24. http://www.mlive.com/news/grand-rapids/index.ssf/2015/11/grand_rapids_mayor_headed_to_p.html.

Van Vuuren, Detlef P., Jae Edmonds, Mikiko Kainuma, Keywan Riahi, Allison Thomson, Kathy Hibbard, George C. Hurtt, et al. 2011. "The Representative Concentration Pathways: An Overview." *Climatic Change* 109 (1–2): 5. https://doi.org/10.1007/s10584-011-0148-z.

Villaraigosa, Mayor Antonio R., Varun Sivaram, and Ron Nichols. 2013. "Powering Los Angeles with Renewable Energy." *Nature Climate Change* 3 (9): 771–775. https://doi.org/10.1038/nclimate1985.

W., Christian. 2017. "Copenhagen Sets New Housing Construction Record." *Copenhagen Post*, February 18. http://cphpost.dk/news/business/copenhagen-sets-new-housing-construction-record.html.

Walther, Gian-Reto, Eric Post, Peter Convey, Annette Menzel, Camille Parmesan, Trevor J. C. Beebee, Jean-Marc Fromentin, Ove Hoegh-Guldberg, and Franz Bairlein. 2002. "Ecological Responses to Recent Climate Change." *Nature* 416 (6879): 389–395. https://doi.org/10.1038/416389a.

Weaver, Timothy. 2016. *Blazing the Neoliberal Trail: Urban Political Development in the United States and the United Kingdom*. Philadelphia: University of Pennsylvania Press.

Westerling, A. L., H. G. Hidalgo, D. R. Cayan, and T. W. Swetnam. 2006. "Warming and Earlier Spring Increase Western U.S. Forest Wildfire Activity." *Science* 313 (5789): 940–943. https://doi.org/10.1126/science.1128834.

Wheeler, Stephen M. 2008. "State and Municipal Climate Change Plans." *Journal of the American Planning Association* 74 (4): 481–449.

World Mayors Council. n.d. "History." http://www.worldmayorscouncil.org/about/history.html.

World Mayors Council on Climate Change. 2010a. "COP President and UNFCCC Executive Secretary Join Forces with Cities." Cancun, Mexico. http://www

.worldmayorscouncil.org/press-room/news-detail/article//cop-president-and-unfccc-executive-secretary-join-forces-with-cities.html.

———. 2010b. "World Mayors Council on Climate Change—Mexico City Pact." 2010. http://www.worldmayorscouncil.org/the-mexico-city-pact.html.

WRI, CAIT. 2014. "Climate Analysis Indicators Tool: WRI's Climate Data Explorer." Washington, DC: World Resources Institute. http://cait2.wri.org.

Wright, Sharon D., and Richard Middleton. 2001. "The 2001 Los Angeles Mayoral Election: Implications for Deracialization and Biracial Coalition Theories." *Politics & Policy* 29 (4): 692–707. https://doi.org/10.1111/j.1747-1346.2001.tb00611.x.

Yates, Douglas T. 1977. *The Ungovernable City: The Politics of Urban Problems and Policy Making*. Cambridge, MA: MIT Press.

Yienger, James, Lizbeth Brown, and Nancy Skinner. 2002. "Experiences of ICLEI's Cities for Climate Protection Campaign (CCP): A Focus on Asia." In *Proceedings of IGES/APN Mega-City Project*. Kitakyushu, Japan.

Young, Oran R. 2008. "Institutions and Environmental Change: The Scientific Legacy of a Decade of IDGEC Research." In *Institutions and Environmental Change: Principal Findings, Applications, and Research Frontiers*, by Oran R. Young, Leslie A. King, and Heike Schroeder, 3–45. Cambridge, MA: MIT Press.

Younger, Margalit, Heather R. Morrow-Almeida, Stephen M. Vindigni, and Andrew L. Dannenberg. 2008. "The Built Environment, Climate Change, and Health: Opportunities for Co-Benefits." *American Journal of Preventive Medicine* 35 (5): 517–526.

Zahniser, David, and Phil Willon. 2010. "Steamed at Villaraigosa, L.A. City Council to Review His Electric Rate Hikes." *Los Angeles Times*, March 24. http://articles.latimes.com/2010/mar/24/local/la-me-dwp-rates24-2010mar24.

Zahran, Sammy, Samuel D. Brody, Arnold Vedlitz, Himanshu Grover, and Caitlyn Miller. 2008. "Vulnerability and Capacity: Explaining Local Commitment to Climate-Change Policy." *Environment and Planning C: Government and Policy* 26: 544–562.

Index

Aggarwala, Rohit, 118, 165
Andrew, Caroline, 178
Anguelovski, Isabelle, 68
Ariella, Marion, 93
auditing legislation, 135–36, 140, 141

Ban Ki-moon, 30
Barber, Benjamin R., 1
Beauregard, Robert A., 178
behavior change, 53, 54, 58, 64, 75, 152
benchmarking legislation, 112, 128, 135–37, 140–41, 144, 151–52, 158, 162
Better Buildings Partnership, 91, 106, 107, 155, 162, 166
Bloomberg, Michael: and catalytic effects of urban climate change mitigation governance, 157–58; as chair of C40, 25–26, 30; on cities as solution to climate change, 37; and congestion pricing, 96; and context of climate change mitigation in New York City, 82; and coordination in institution building, 118, 119; and determining opportunities for intervention, 59; on financial health of New York City, 83; and Global Compact of Mayors, 31; on importance of measurements, 135, 150; and Mayor's Carbon Challenge, 95; and policy agendas for reducing GHG emissions, 93; as spokesperson at United Nations, 34
Bradford, Neil, 177
branding, 34–35
buildings: and energy regulation in New York City, 94–95, 151; retrofitting, in New York City, 81–82; retrofitting, in Toronto, 105–7
Bulkeley, Harriet, 42, 61–62, 66
Burney, Jerrel, 165
Bush, George W., 23
Byrd–Hegel Resolution, 23
Bystryn, Marcia, 82–83

C40 Cities Climate Leadership Group, 25–26, 28, 29, 30, 31, 40, 72, 185n5
California, 41. *See also* Los Angeles
California Global Warming Solutions Act (2006), 88

Canada: and Kyoto Protocol, 23; and Municipal Leaders Resolution on Climate Change, 185n4; neoliberal reforms in, 3; power of city governments in, 5. *See also* Toronto
capacity building, 71–73, 134–42, 144–45, 165, 171–72
cap and trade programs, 24, 61, 83, 88, 92, 108, 176
carbon Cities Climate Registry (cCCR), 31
carbon credits, 108, 138, 156
Carmin, JoAnn, 68
Castán Broto, Vanessa, 46, 145
Change is in the Air, 104, 121
cities: contributions to GHG emissions, 19–20; vulnerability to impacts of climate change, 19, 37–38. *See also* city branding; city government emissions; city governments
Cities for Climate Protection (CCP), 27
Cities Initiative, The, 26
city branding, 34–35
City Energy Project, 158
city government emissions: in Los Angeles, 153–54; in New York City, 150; in Toronto, 107–8
city governments: as agents of change, 1–2, 3, 6, 17, 29–32, 165–67; catalytic effects of urban climate change mitigation governance, 156–63; and challenge of climate change mitigation, 50–57; and climate change mitigation, 3–5, 174; in contemporary urban governance settings, 11–12; contributions to study of climate change mitigation and, 11–14; emergence of response to climate change, 25–32; limitations of, 3, 6; multidimensional evaluation of actions of, 73–76; optimistic portrayal of, 2–3; policy challenges and, 178–79; reasons for response to climate change, 32–39; and repowering cities, 5–8; research design and methods of study, 8–11; role in urban governance, 47–50; and scaling up climate change mitigation efforts, 168–74; scholarship on, 12–13; sharing and comparing experiences between, 168; unique contexts of, 79. *See also* governing strategies; urban governance

209

210   INDEX

City of Toronto Act (2006), 89–90
Clean Heat, 140
Clean Heat Accelerator, 140
CLEAN LA Solar Coalition, 127–28
climate change: causes of, 18; cities' vulnerability to impacts of, 19, 37–38; emergence of urban response to, 25–32; global response to, 21–25; impact of, 18–19; limits and tradeoffs of local response to, 174–77; reasons for urban response to, 32–39. *See also* climate change mitigation; greenhouse gas (GHG) emissions; progress on climate change mitigation
climate change mitigation: approaches to, 185n3; catalytic effects of urban governance, 156–63; cities' contributions to, 32–33, 174; city governments and challenge of, 50–57; city governments' impact on, 165–67; co-benefits of, 35–37; complexity of urban, 51–52, 55; context of, in Los Angeles, 84–88; context of, in New York, 81–84; context of, in Toronto, 88–92; contributions to study of city governments and, 11–14; evaluation of outcomes and consequences of efforts in, 8; factors and barriers influencing success in, 41–44; gap between urban ambition and progress in, 39–41; governing modes and, 61–66; local benefits of, 35–37; multidimensional evaluation of, 73–76; and navigating complex policy landscape, 44–45; policies and programs for energy, transportation, and waste management sectors, 51f; policy challenges concerning, 177–79; public support for, 37–38; and repowering cities, 5–8; research design and methods of study on, 8–11; scale of city involvement in, 185n5; scaling up efforts in, 168–74; scholarship on, 12–13; sector choice and, 58–61; steps for, 20–21; tailored policy agendas for, 111–13, 167; uncertainty of urban, 52–55, 67–72; undertaken by city governments, 3–5; United States as obstacle to, 22. *See also* progress on climate change mitigation
Climate Group, The, 28
Clinton Climate Initiative (CCI), 28
coal, 149, 154, 155, 161
coalition building, 69–71, 125–34, 142–44, 145, 172–74
Conference of the Parties, 23
congestion pricing, 95–96, 134
Conservation Law Foundation, 133
Consolidated Edison (ConEdison), 82, 97, 152, 186n4

Constantinides, Costa, 123
Copenhagen, Denmark, 35

Daley, Richard, 26
data collection and sharing, 31, 72, 138, 144, 171–72
Data LA initiative, 160
de Blasio, Bill: and coalition building, 127; and determining opportunities for intervention, 59; and long-term institution building, 123; and Mayor's Carbon Challenge, 95; and policy agendas for reducing GHG emissions, 93–94; supports renewable energy supplies, 97, 98; and task force model, 157
deep lake water cooling system, 91, 155, 190n17

economic conditions, impact on city priorities, 61
Empowerment Zone programs, 55
enabling mode of governance, 64, 65
Energy Efficiency and Conservation Block Grant program, 30–31
Energy Efficiency Office (Toronto), 91
Energy Independence and Security Act (2007), 84
energy rate increases, 129–32, 159
Energy Savings Performance Standards, 109
energy services, sources, and systems: impact on city priorities, 60; Los Angeles policy agendas concerning, 100–102; New York City policy agendas concerning, 97–98
entrepreneurial cities, 34–35
Environment LA, 120
European Covenant of Mayors, 28
Existing Building Energy and Water Efficiency Ordinance, 128, 136

federal and subnational governments: action and inaction of, 38–39; city governments' impact on, 75–76
Federation of Canadian Municipalities' Partners for Climate Protection program, 26
Figueres, Christiana, 30
financing: Better Buildings Partnership, 91, 106, 107, 155, 162, 166; for energy efficiency retrofits, 59, 105; Energy Savings Performance Standards, 109; for GHG emission reductions, 56; Green Building Financing, 141; Home Energy Loan Program (HELP), 106–7, 128–29, 155; provided by Toronto Atmospheric Fund, 92, 109; transportation, in Los Angeles, 86; transportation, in New

York City, 82. *See also* Green Building Financing
Ford, Doug, 91
Ford, Rob, 91, 104, 110, 124, 129

Gaebler, Ted, 49
Garcetti, Eric: and capacity building, 136; and catalytic effects of urban climate change mitigation governance, 160; and institution building, 119, 120, 125, 142; and Mayors National Climate Action Agenda, 30; and policy agendas for reducing GHG emissions, 99, 100; as roadblock to environmental initiatives, 146
Gennaro, James, 123
Global Covenant of Mayors for Climate & Energy, 31, 35
Global Protocol for Community-Scale GHG Emissions Inventories (GPC), 31, 85, 189n1
Gold, Mark, 99
Good, Kristin, 70
governing strategies, 116–17; capacity building, 71–73, 134–42, 144–45, 171–72; challenges and varieties of mobilization, 142–47; coalition building, 69–71, 125–34, 142–44, 145, 172–74; employed by case studies, 166; institution building, 68–69, 117–25, 142, 145, 169–71; mobilizing actors and resources, 66–73; and policy agendas for reducing GHG emissions, 112, 113t; and repowering cities, 7; sharing and comparing experiences regarding, 168; in urban governance policy agenda, 61–66. *See also* city governments; urban governance
government capacities, 135–39
Grand Rapids, Michigan, 33–34
green bills, 126
Green Building Financing, 141
Green Buildings Technical Working Group, 157
Green Climate Cities (GCC) program, 26, 27–28, 40, 72, 185n5
Green Codes Task Force, 126, 157
Greener, Greater Buildings Plan (GGBP), 95, 135, 140, 152
greenhouse gas (GHG) emissions: cities' contributions to, 19–20; city governments and measurement of, 31; clarification of boundaries of, 52–53; federal limitations on, 83–84; increase in, 18–19; inventorying, 52; linking reductions to city government actions, 166; local benefits of reducing, 35–37; Los Angeles's policy agenda for reducing, 98–103; measuring change in, 73–76; New York City's policy agenda for reducing, 92–98; primary routes taken to reduce, 113t; profile of New York City, 81; progress on reduction of, 149–56; reducing, 20–21; reduction of Toronto's government-produced, 107–8; Toronto's policy agenda for reducing, 104–11; tracking, 72, 74. *See also* climate change mitigation; progress on climate change mitigation

Green LA: An Action Plan to Lead the Nation in Fighting Global Warming, 88, 99, 120, 124, 153, 160, 188n21
Green LA coalition, 99, 187n44
Green Light New York, 139
GreeNYC, 139
Gronewold, Nathanial, 151
Gross, Jill Simone, 46

Hahn, James, 187n42
Hansen, Jim, 21
Heartwell, George, 1, 33–34
heating fuels, 60, 140, 150
Hidalgo, Anne, 25
High Rise Retrofit Support (HiRIS), 106–7
Hoffmann, Matthew J., 24–25, 32
Home Energy Loan Program (HELP), 106–7, 128–29, 155

ICMA Local Government Sustainability Survey (2010), 41
Imbroscio, David, 178
income inequality, 173
Independent Electricity System Operator (IESO), 89
Indiana, 40
institution building, 68–69, 117–25, 142, 145, 169–71
Integrated Power Resource Plan, 154
Integrated Project Delivery, 109
intercity networks, 39, 51–52, 72–73, 185n5. *See also* C40 Cities Climate Leadership Group; Green Climate Cities (GCC) program)
Intergovernmental Panel on Climate Change (IPCC), 18
International Brotherhood of Electrical Workers Local 18, 87, 130
International Council on Local Environmental Initiatives (ICLEI), 26–28, 39–40. *See also* Green Climate Cities (GCC) program

Jacobs, Alan, 186n2
Jacot, David, 102, 116–17
Jensen, Frank, 35
job creation, 36, 66, 70, 100–101, 129, 172

Katz, Bruce, 70
Kern, Kristine, 42
Kerry, John, 1
Klijn, Erik-Hans, 52, 54
Koppenjan, Joop, 52, 54
Krause, Rachel M., 40
Kyoto Protocol, 22–23, 28

labor unions, 87, 130
Livingstone, Ken, 28
Local Law 84, 135
Local Law 87, 135
London Climate Change Agency, 162
long-term institution building, 122–25
Los Angeles: capacity building in, 136, 137–38, 141, 144, 145; catalytic effects of climate change mitigation governance in, 159–61; climate change commitments from, 77t; climate change mitigation efforts of, 8–9; climate change plan of, 36; coalition building in, 127–28, 129–33, 143–44, 145; context of climate change mitigation in, 84–88; contribution of energy, transportation, and waste to GHG emissions in, 10f; destinations of New York and Toronto and, 163t; energy supply of, 60; framing of climate change programs in, 70; governing strategies employed by, 166; green building codes adopted by, 39; institution building in, 119–21, 124–25, 142, 145; plans for reducing GHG emissions in, 29; policy agenda for reducing GHG emissions in, 98–103, 111, 112; political commitment in, 146; political tensions in, 171; primary routes taken to reduce GHG emissions in, 113t; progress in reduction of GHG in, 149, 152–54; role of context in shaping policy agendas of, 113–15; steering strategies for climate change mitigation in, 145t; steering strategies for reducing uncertainties in climate change mitigation, 67f; as study case, 11, 77–78; summary of terrain, 81t
Los Angeles County Metropolitan Transportation Authority (LA Metro), 86
Los Angeles Department of Water and Power (LADWP), 84–87, 101–2, 120–21, 128–32, 136–38, 143, 144, 154, 159
Los Angeles Neighborhood Councils, 159
low-emission public housing, 109

Mark-Viverito, Melissa, 123
mayors: and determining opportunities for intervention, 59; influence of, 82–83; of Los Angeles, 86–87; of New York City, 186n7. *See also* Bloomberg, Michael; de Blasio, Bill; Garcetti, Eric; Villaraigosa, Antonio
Mayors National Climate Action Agenda, 30, 160
Measure J, 103
Measure R, 102, 125, 133, 166
Measure RRR, 132
methane capture, 107, 149, 151, 155
Metropolitan Transportation Authority (MTA), 82, 134
Mexico City Pact, 29–30
Miller, David, 25, 90, 91, 104, 121
mobilization: of political resources, 56; of technical resources, 56–57; through governing strategies, 66–73
Montreal Protocol on Substances That Deplete the Ozone Layer (1987), 21
Moving America Faster bill (2011), 103
Municipal Leaders Resolution on Climate Change, 185n4

natural gas, in New York City, 97, 150–51
neighborhood councils, in Los Angeles, 159
New Localism, 1
New York City: capacity building in, 135–36, 139–41, 144, 145; catalytic effects of climate change mitigation governance in, 157–59; climate change commitments from, 77t; climate change mitigation efforts of, 8–9; coalition building in, 126–27, 134, 143, 145; context of climate change mitigation in, 81–84; contribution of energy, transportation, and waste to GHG emissions in, 10f; destinations of Los Angeles and Toronto and, 163t; energy supply of, 60; as example of self-governing mode of governance, 63; governing strategies employed by, 166; institution building in, 118–19, 122–23, 142, 145; plans for reducing GHG emissions, 29; policy agenda for reducing GHG emissions, 92–98, 111, 112; political commitment in, 146; primary routes taken to reduce GHG emissions in, 113t; progress in reduction of GHG in, 149, 150–52; responsibilities of mayor of, 186n7; role of context in shaping policy agendas of, 113–15; steering strategies for climate change mitigation, 145t; steering strategies for reducing uncertainties in climate change mitigation, 67f; as study case, 11, 77–78; summary of terrain, 81t; switches to Global Protocol for Community-Scale GHG Emissions Inventory, 189n1

New York City's Roadmap to 80×50, 159
New York Energy Efficiency Corporation, 141
Nickels, Greg, 28, 39
Nowak, Jeremy, 70

Obama, Barack, 23–24, 38
Office of Long Term Planning and Sustainability (OLTPS), 118–19
Office of Public Accountability, 132
One City, Build to Last: Transforming New York City's Buildings for a Low-Carbon Future, 93–94, 127, 157
Ontario Power Generation, 89
Osborne, David, 49

Paris Climate Agreement, 23–24, 31
Partners for Climate Protection (PCP), 27, 28, 39–40
Pasztor, Janos, 30
Perry, Jan, 120
Pierre, Jon, 49, 175, 178
pLan, 99, 160
PlaNYC 2030: A Greener Greater New York, 82, 92–93, 118–19, 122–23, 157–58
policy agendas: of Los Angeles for reducing GHG emissions, 98–103; of New York City for reducing GHG emissions, 92–98; in repowering cities framework, 7, 57–66; role of context in shaping, 113–15; tailored, 111–13, 167; of Toronto for reducing GHG emissions, 104–11
policy challenges, 177–79
policy landscape, navigating, 44–45
political resources, mobilization and exchange of, 56
Portney, Kent E., 68
Port of Los Angeles, 85
private finance, 175–76
program integrity, 66
Progressive Los Angeles Network (PLAN), 98–99
progress on climate change mitigation, 148–49, 163–64; catalytic effects of urban climate change mitigation governance, 156–63; reductions in GHG emissions, 149–56
Proposition 13, 87
provisioning mode of governance, 64
public health, 172
public housing, low-emission, in Toronto, 109
public support, 37–38. *See also* voter support

Quinn, Christine, 93, 123

Ratepayer Advocate, 132, 143, 144, 159
real estate industry, and coalition building in New York City, 126
refundable debt model, 106–7
regulating mode of governance, 63
renewable energy: in Los Angeles, 100–101, 137–38, 154, 160–61; in New York City, 97
Reporting on Energy and Water Use Regulation, 137
repowering cities framework: application of, 76–78; governing strategies, 66–73; multidimensional evaluation of impacts, 73–76; policy agendas, 57–66; policy challenges and, 177–79; relevance for other cities, 167–68. *See also* governing strategies; policy agendas; urban governance
Retrofit Accelerator, 140
retrofitting buildings, 81–82, 105–7
Rich, Michael J., 55, 66
Robert Cooke Co-op, 109
Rockefeller 100 Resilient Cities, 73

Sadik-Khan, Janette, 119
Scope 1 emissions, 53
Scope 2 emissions, 53
Scope 3 emissions, 53
sectors, in urban governance policy agenda, 58–61
self-governing mode of governance, 62–63, 65
Selin, Henrik, 75
Smart Track, 110
Smith, Greg, 130
social inequality, 178
SolarCity, 160–61
Solar Incentive Program, 101, 137, 144, 160
Solar LA plan, 100–101, 131
solar power, 97, 100–101, 137–38, 152, 160–61
Southern California Gas Company, 121
stakeholder capacities, 139–42
stakeholder coalitions, 125–29, 143, 159
state governments, city governments' impact on, 75–76
state retrenchment, 3
steering, 49–50, 145t
Stoker, Robert P., 55, 67
Stone, Clarence N., 66
strategic urbanism, 28
Sustainable Energy Funds, 105–6
Sustainable Energy Plan Financing, 106

taxis, 96, 152
technical resources, mobilization and exchange of, 56–57